Tomorrow's People

SUSAN GREENFIELD

Tomorrow's People

HOW 21ST-CENTURY
TECHNOLOGY IS CHANGING
THE WAY WE THINK AND FEEL

ALLEN LANE
an imprint of
PENGUIN BOOKS

ALLEN LANE

Published by the Penguin Group
Penguin Books Ltd, 80 Strand, London WC2R 0RL, England
Penguin Putnam Inc., 375 Hudson Street, New York, New York 10014, USA
Penguin Books Australia Ltd, 250 Camberwell Road, Camberwell, Victoria 3124, Australia
Penguin Books Canada Ltd, 10 Alcorn Avenue, Toronto, Ontario, Canada M4V 3B2
Penguin Books India (P) Ltd, 11 Community Centre, Panchsheel Park, New Delhi – 110 017, India
Penguin Books (NZ) Ltd, Cnr Rosedale and Airborne Roads, Albany, Auckland, New Zealand
Penguin Books (South Africa) (Pty) Ltd, 24 Sturdee Avenue, Rosebank 2196, South Africa

Penguin Books Ltd, Registered Offices: 80 Strand, London WC2R 0RL, England

www.penguin.com

First published 2003
1

Set in 10.5/14 pt PostScript Linotype Sabon
Typeset by Rowland Phototypesetting Ltd, Bury St Edmunds, Suffolk
Printed in England by Clays Ltd, St Ives plc

ISBN 0–713–99631–5

τιγγω

Contents

Preface

This book should really have been a novel. Like many, I had often fantasized about conveying a stream of thoughts and insight effortlessly through my fingertips, of telling a story where I myself didn't know the outcome, and writing of characters who developed independent lives and minds of their own. Novel-writing had appeal not just as a means of letting one's mind range free but also because it could do so without all the checks and balances of argument and reference, the meticulous research and fact-checking, that characterizes non-fiction books of the type that I had previously written on the brain. Indeed, such unfettered literary abandonment was just the type of activity I was looking forward to on a quiet beach holiday in the Caribbean during the Christmas break, a few years ago.

But after several days, with a pitiful page or two recriminating me for squandering precious free time for nothing, I had to admit that I was bored with my own efforts. When the author herself finds her narrative pedestrian, her characters clichéd and her dialogue like a long-lost episode of *The Woodentops* it is time to think again. I abandoned the project and returned to London irritated and frustrated. The problem was that, aside from the thwarted glamour of writing a novel, I did have an interesting idea I really wanted to develop.

Although my day job is research neuroscience, with a primary interest in neuronal mechanisms underlying neurodegenerative disorders such as Alzheimer's and Parkinson's diseases, I have long been fascinated by the broader and still more slippery question of how the physical brain generates the subjective inner state that we call consciousness. I had, in 1995, attempted a neuroscientific contribution to this question with *Journey to the Centers of the Mind*, followed in

2000 with *The Private Life of the Brain*. Needless to say, the 'hard problem' of turning the water of physiology into the wine of phenomenology remained unresolved in both cases. But thinking about the issue over and over did prompt a kind of meta-question: what would it be like if the problem was solved after all? What kinds of lives would we all be living? This is what the novel would have explored, and for good measure through the eyes of a brilliant and beautiful heroine, a female neuroscientist . . .

But reality had intervened and demonstrated with cruel clarity that character development, pace of narrative and gripping dialogue are far from easy. As I was bemoaning my shortcomings over lunch with Stefan McGrath of Penguin, who had been my hugely supportive editor for *The Private Life of the Brain*, he cut me short with a simple question: what was so alluring about what I had been trying to do? The short answer was that I had been yearning to use my imagination.

As a research scientist I get to speculate from time to time, planning a new set of experiments or trying to interpret a puzzling finding. But speculation unsubstantiated by published data holds no currency. One is always on guard against creating hypothetical scenarios that might account for a result but take too much for granted empirically or disregard existing dogma too uncritically. Yet still, at the end of the day, in the bar or restaurant with colleagues, it can be enormously rewarding intellectually – and indeed helpful back at the bench – to explore the big, empirically intractable questions, to put detailed findings into a wider context, above all to ask 'what if . . . ?'

So, suggested Stefan, I could indulge my imagination and ask this question, without fretting that I was no Jane Austen: I could write a non-fiction book on the future of our brains or, more accurately, our minds. The idea, once voiced, seemed dead right. During the previous few years I had been reading a little about the new science of nanotechnology and had appeared on the panel of the TV programme *Futurewatch*, as well as being very impressed by astrophysicist Michio Kaku's *Visions*, which documents 'how science will revolutionize the 21st century and beyond'. I had also happened to hear a lecture by Ian Pearson, a futurologist from British Telecom, and had been chilled and excited in equal measure as he unfolded his predictions of a lifestyle dominated by a highly personalized yet intrusive IT.

If these possibilities came to pass, I wondered, what impact might they have on the brain? After all, the 'plasticity' of the human brain – how it can reflect individual experience – is widely acknowledged. How might all this technology change our outlook? Moreover, in my excursions into the neuroscience of consciousness I had explored how the personalized brain could, due to mental illness, drugs, dreaming or fast activity, revert to a passive state of raw subjective sensation, a little like a small child or a non-human animal. The new technologies, I came to realize, might also have the effect of putting us into a passive, sensory-laden state where our personalized brains – our minds – become less relevant.

The central issue here, then, isn't the new technologies themselves (on which there exists already a wide range of excellent books, see 'Further Reading', pp. 273–6); instead the question is how the advances in science might change our thoughts, feelings and personalities. As such, this book could not be an exhaustive survey of every aspect of the future – for example, I have left out entirely the possibility of space travel and extraterrestrial life. But by the same token, the central question of how our minds might be transformed does demand a wide sweep of both the physical and biomedical sciences, rather than a focus on just one discipline or topic. I trust therefore that you, the reader, will not feel caught between the two stools of a non-comprehensive survey and an insufficiently detailed account. The brief, as I saw it, really did need a compromise.

In fact, as I travelled further into the chapters it became clear that we would have to touch on important implications relating not just to the human mind per se but also to economics, politics and, indeed, the state of the world. Perhaps almost inevitably, I found that I was closing the book with a chapter on terrorism. I am writing this preface in very unusual times, on the very day when, around the world, there are demonstrations against war on Iraq. I would hope and expect that by the time this book is published that particular situation will have been resolved, but I fear that terrorism in general will not be regarded as passé. It will become a basic factor in our lives in the future, and as such raises vitally important questions concerning freedom of thought, free will, and hence the nature of individuality. So I find I have ended up writing in celebration of the individual human mind, and making a desperate and deadly serious plea that it will be preserved into the future; this end point seems a long way from the stereotyped science-starlet of the original, aborted novel.

Such journeys cannot be made alone. Because of the wide-ranging nature of the material, I needed advice on specialist areas outside my own expertise, and I am deeply grateful to various colleagues who not only fielded questions but actually read early drafts of the MS: Professor Igor Aleksander (Department of Electrical and Electronic Engineering, Imperial College, London); Professor Peter Atkins (Department of Physical Chemistry, Oxford University); Professor Guy Claxton (Graduate School of Education, University of Bristol); the world-class architect Sir Terry Farrell; Professor Bob Williamson (Murdoch Children's Research Institute, Melbourne). In addition I am indebted to Katinka Matson of Brockman Inc., New York, for all her support and advice, and of course Stefan McGrath of Penguin, without whom the whole project would never have happened; I would also like to thank Ben Gribbin of Penguin for meticulous and very valuable line-editing. There are, however, two more colleagues who have proved indispensable in the writing of this book, from its inception to its final delivery. Milla Harrison has provided unflagging support in sourcing and providing the very latest material and information on all the subjects covered; I have yet to work with anyone more willing, efficient and professional. Well, perhaps there is just one: my assistant Viv Pearson, who has not only given 110 per cent secretarial assistance but who, more importantly still, has also been a real friend throughout.

Susan Greenfield, Oxford, February 2003

I

The Future: What is the problem?

Look through an old album of sepia photographs from the early 1900s. There they are, our forebears, most usually posed in front of some cardboard Arcadian scene, doomed to manual or social drudgery and a rigid code of conduct and thought. Those placid, distant faces stare into a world, invisible and unknowable to us, of toothache, outside privies, stale sweat and certainty. 'The past is a foreign country,' mused L. P. Hartley in *The Go-Between*, 'they do things differently there.' Yet the mid-20th-century British prime minister Harold Macmillan, looking back over a long life to his Victorian childhood, once reminisced that the great watchword of the turn of the century was 'progress'. Progress – social, economic and above all scientific – was perceived as just that, the forward march of the human intellect, from which we would reap only benefits. And progress came from science.

In the 1950s the scientist knew everything. He (always he) was characterized in television advertisements as the white-coated authority, condescending to endorse 'scientifically' the latest washing powder. The very fact that there was television at all transformed not only people's lives but also the way they viewed the world beyond the confines of their own community. The chirpy, capped, short-trousered schoolboy of that era, voraciously swotting up endless facts that 'every schoolboy knows', was fascinated by the technological marvels of the Festival of Britain and the new world that science was making possible. Meanwhile penicillin was rescuing many from misery and early death, whilst the contraceptive pill, no longer just a pipe dream, was about to revolutionize the outlook of, and for, women.

But the 20th century has surely taught us, among much else, that everything comes with a price; every schoolchild now knows that scientific and technological advances have colossal potential for both good and evil. Although the public have been aware, ever since Hiroshima, of the need to try to understand the implications of new scientific discoveries, it has only been in the last few decades of the previous century that the alarm bells have grown deafening. GM foods, mad cow disease and brain-scrambling mobile phones have compelled the most ostrich-like technophobe to question what might be happening in the remote and rarefied stratosphere of the laboratory. For science is increasingly not just on our minds but at the heart of our lives, encroaching upon everything that we hold dear: nutrition, reproduction, the climate, communication and education . . . The impact of science and technology on our existence, in the future, is no longer a whimsical excursion into science fiction.

Those sci-fi images of yesteryear now have an enchantingly amateurish glow. The Daleks in pursuit of Dr Who, the politically correct crew in *Star Trek* – even that ultimate icon, from Stanley Kubrick's film *2001*, the psychopathic computer, HAL – are as far-fetched and unthreatening as the tin-foil outfits and staccato jerks of the marionettes in *Thunderbirds*. The human and humanoid characters, in most cases, think and act like we do. They have similar sets of values and expectations, and the bulk of the appeal depends on a good guys/bad guys plot. And that is how most people used to see the future – not chasing bandits around the galaxy so much as still being human in a world of souped-up, high-tech gadgetry – a gadgetry perhaps of interest to some anorak-kitted nerds, but for the majority of us reasonable everyday folk to be taken in our stride.

But now we face a future where science could actually change everyday life any day soon; many think such transformations are already under way. Yet there are some – let's dub them, without much originality, The Cynics – who do not see any point in dusting down the crystal ball. The chances are, glancing at the track records of our predecessors, that pretty much any prediction anyone makes now will be either impractical or uninspired.

Moreover, just because a technology is up and running doesn't mean to say it will become central to the daily grind. One late-19th-century prediction of the future, for example, was that everyone would travel around in hot-air balloons. And on the other hand, unknown, unimaginable technology can catch us unawares: a picture of a domestic scene 'in the future' drawn back in the 1950s shows all manner of gleaming appliances, but no computers, let alone anyone surfing the web. Even a glimmer of the priming technology just wasn't part of normal existence; it would have been a fairly impressive intellectual leap to conceptualize our 50-emails-a-day lifestyle from the standing start of clunky, expensive and essentially mechanical computers whirring and churning in their remote rarity in custom-made rooms of their own. And I remember a summer afternoon in the 1970s, lounging after a heavy lunch on a lawn with friends, when someone, a physicist, first mentioned the microchip – he prophesied that 'it will change all our lives'. The rest of us hadn't the vaguest idea what he was talking about.

The problem with thinking about the future, shrug The Cynics complacently, is that it is impossible to predict the big new scientific advances that underpin serious technological progress; meanwhile, how easy to be distracted by high-tech toys, the latest variation on an existing theme, amusing enough for escapist science fiction but not sufficiently innovative to restructure our entire existence and our seemingly impregnable mindset. Yet, as physicist Michio Kaku points out, the problem with extrapolating the future in the past – as with the hot-air balloon mass transport system – is that it hasn't been the scientists themselves making the predictions. Now they are in a very strong position to do so.

However, The Cynics have long placed a trip wire on the track of human progress, even when scientists have indulged in flights of fancy. They laughed at Christopher Columbus, derided Galileo, scoffed at Darwin and sneered at Freud. A curious feature of The Cynic's attitude is that he (and again it usually is he) thinks that science is on his side, backing up his sane voice of reason against the fantastic. In 1903 a *New York Times* editorial glibly wrote off Langley's attempts at flight: 'We hope that Professor Langley will not put his substantial greatness as a scientist in further peril by continuing to waste his time, and the money involved, in further airship experiments. Life is short, and he is capable of services to humanity incomparably greater than can be expected to result from trying to fly . . .' And a few decades later, in 1936, when technology had become much more part of life, Charles

Lindbergh wrote to Harry Guggenheim of Robert Goddard's rocket research: 'I would much prefer to have Goddard interested in real scientific development than to have him primarily interested in more spectacular achievements which are of less real value.'

Even now one of the most popular quotes for after-dinner speeches has to be the famous prediction of Thomas Watson, Chairman of IBM, in 1943: 'I think there is a world market for maybe five computers.' And if you had suggested to our 1950s schoolboy that one day his, or her, 21st-century counterpart would have no idea what a slide rule was, or what log tables were all about, they would have thought you utterly crazy.

But it still does not follow that *this* time, *this* century should be any different, in terms of the revolutions in science and technology that come and go. Yes, as we shall see, we may well have the technology for a disease-free, hunger-free and even work-free existence. But then, too, the values, fears and hopes engendered in a chilly, smelly cottage on a bleak hillside would have produced an outlook very different from one based on a 20th-century upbringing in a centrally heated suburbia shimmering with shiny, chrome appliances and unforgiving neon lights. Yet we still have the same human brains as our very early ancestors, who stumbled uncomprehendingly around on the savannah some 100,000 years ago.

For the first time, however, our brains and bodies might be directly modified by electronic interfaces. For a second group, The Technophiles, such a prospect is welcome. The electrical engineer Kevin Warwick, for one, would welcome the prospect of heightened senses, sensations and muscle power that being a cyborg might bring – as we will see later. And cyber-guru Ray Kurzweil is gung-ho for the intimate embrace of silicon:

There is a clear incentive to go down this path. Given a choice, people will prefer to keep their bones from crumbling, their skin supple, their life systems strong and vital. Improving our lives through neural implants on the mental level, and nanotechnology-enhanced bodies on the physical level, will be popular and compelling. It is another one of those slippery slopes – there is no obvious place to stop this progression until the human race has largely replaced the brains and bodies that evolution first provided.

Both Warwick and Kurzweil, not to mention other intellectual luminaries such as Marvin Minsky and Igor Aleksander, along with various futurologists such as Ian Pearson and Hans Moravec, all take it as read that another feature of future life will be conscious machines. Kurzweil's message is that our only future as a species will be to merge intimately with our technology: if you can't beat the robots, join them. So imagine a spectrum of beings, from pure carbon-based (as we humans are now) through the cyborg silicon-carbon hybrids that we could become to the ultimate – the vastly superior thinking silicon systems that will be Masters (and again they will have to be male) of the Universe.

It was actually because he was eavesdropping on a discussion between Kurzweil and the philosopher John Searle, concerning the very question of computer consciousness, that the co-founder and Chief Scientist of Sun Microsystems, Bill Joy, began to feel anxious about the direction in which future technology was heading. As an undisputed techno-mandarin, Joy created an enormous stir when he wrote of his urgent concern in the magazine *Wired*, in April 2000, in an article titled 'Why the future doesn't need us':

The 21st-century technologies – genetics, nanotechnology, and robotics – are so powerful that they can spawn whole new classes of accidents and abuses. Most dangerously, for the first time, these accidents and abuses are widely within the reach of individuals and small groups. They will not require large facilities or rare raw materials. Knowledge alone will enable the use of them.

True, a critical difference between the technology of the 21st-century genetics, nanotechnology and robotics and that of the previous 100 years – darkening as they were with nuclear, biological and chemical doom – is that now it is no longer necessary to take over large facilities or access rare raw materials. Yet an even bigger change in the technology of the future, compared to that of the past, is that a nuclear bomb, though hideous in its potential, cannot self-replicate; but something that might – nanorobots – could soon be taking over the planet.

Just browse a few websites that are devoted to 'problems of preserving our civilization'. One worry, you will read, is that the manipulation of matter at the level of atoms, the nanotechnology that promises to be 'the manufacturing industry of the 21st century', will bring a new enemy – robots scaled down to the billionth of a metre that the nanolevel mandates, minuscule serfs who are focused on assembling copies of themselves. What might happen, one website asks, if such prolific yet single-minded operatives fell into the hands of even a lone terrorist? But then, of course, intelligent robots do not have to be small to be evil – just much cleverer than us. Common-or-garden human-sized machines might also soon be able to self-assemble, and, more importantly, to think autonomously.

Bill Joy had never thought of machines heretofore as having the ability to 'think'; now he is worried that they will, and in so doing lead us into a technology that may replace our species. He worries that humans will become so dependent on machines that we will let machines make decisions. And because these machines will be so much better than humans at working out the best course of action, soon we will capitulate entirely. Joy argues that, in any case, the problems will soon be so complex that humans will be incapable of grasping them. Considering that, in addition to greater mental prowess, these silicon masterminds will have no need to sleep in, nor to hang out in bars, they will soon be way ahead of us, treating us as a lower species destined, as one website warns, to be 'used as domestic animals' or even 'kept in zoos'.

Kevin Warwick's predictions are similarly ominous: 'With intelligent machines we will not get a second chance. Once the first powerful machine, with an intelligence similar to that of a human, is switched on, we will most likely not get the opportunity to switch it back off again. We will have started a time bomb ticking on the human race, and we will be unable to switch it off.'

Equally nightmarish would be an elite minority of humans commanding large systems of machines, whilst the masses languish redundant. Either the elite will simply destroy this useless press of humanity or, in a more benign mood, generously brainwash them so that they give up reproducing and eventually make themselves extinct – it would be kindest to ensure that at all times the masses are universally content. They will be happy, but not free. It is a disturbing thought that these are the views of the Unabomber, Theodore Kaczynski; though he was obviously criminally insane, and no one would for a

moment condone his actions, still Joy felt compelled to confront the
sentiment that 'as we are downloaded into our own technology, our
humanity will be lost'.

The coming Age of IT, then, offers a raft of possibilities from
conscious automata to self-assembling autocrats to carbon-silicon
hybrids. Extreme though such possibilities might seem, especially to
The Cynics, it is very likely that a more modest version of carbon-
silicon interfacing will feature in 21st-century life before too long.
Soon computers will be invisible and ubiquitous – if not actually inside
our bodies and brains then sprinkled throughout our clothes, in our
spectacles and watches, and converting the most unlikely inanimate
objects into 'smart' interactive gadgets.

The real problem is not what is technically feasible but the extent to
which what is technically feasible can change our values. The gadgets
of applied technology are the direct consequences of the big scientific
breakthroughs of the previous century, and promise any day now to
influence, with unprecedented intimacy, the previously independent,
isolated inner world of the human mind. Yet this widespread availability of modern technology is, for some, a loud enough wake-up call
for us to re-evaluate our priorities as a society. Bill Joy again: 'I think
it is no exaggeration to say that we are on the cusp of the further
perfection of extreme evil, an evil whose possibility spreads well
beyond that which weapons of mass destruction bequeathed to the
nation-states, on to a surprising and terrible empowerment of extreme
individuals.'

But of course not all of this third group, The Technophobes, are
scientists. Not surprisingly, and indeed more typically, non-scientists'
fears are usually grounded in a more romantic view of life, but the
fears are there nonetheless. In his Reith Lecture in 2000 Prince Charles
summed up the worries of many: 'If literally nothing is held sacred
anymore . . . what is there to prevent us treating our entire world as
some "great laboratory of life", with potentially disastrous long-term
consequences?'

It may be a little unfair, and certainly incautious, to write off this
type of view as simply that of latter-day Luddites, striving in vain to
hold back progress with a misconceived vision of some golden bygone
age when humans adhered to a Rousseau-like natural nobility, and

no one died in childbirth, suffered poor housing, worked at mind-numbing manual tasks or froze to death ... It's just that for many there is a very real fear that science, and the technology that it has spawned, have outpaced the checks and balances we need for society to survive – indeed for life as we know it to continue at all.

In our growing knowledge of life, in biology, the trend for science to be slipping out of control appears already to be gaining an ever faster pace. The rigid hierarchy of a society segregated by biochemical and genetic manipulation, from intellectual 'alphas' down to 'epsilons' who operate the lifts, portrayed by Aldous Huxley in *Brave New World*, is now seen as a real future threat by many. Predictably, a morass of websites express serious concerns over genetics, for example: 'The path is open, by-passing the natural evolution, to design unusual creatures – from fairly useful to imagination-striking monsters.'

And we might well end up with 'designer' babies, potential geniuses or highly obedient and tough soldiers. But manipulation of genes allows further possibilities too; offset against the benefits of gene therapy and new types of medication and diagnostics, there are clones, artificial genes, germ-line engineering, and the tricky relationship of genetic profiling to insurance premiums and job applications. In any event, for The Technophobes, the question of basic survival seems far from certain; according to Bill Joy, the philosopher John Leslie puts the risk of human extinction at 30 per cent at the least. And the astronomer Martin Rees, in his latest book, *Our Final Century*, rates the chance as no better than odds on that civilization will avoid a catastrophic setback.

No one could really disagree with Aristotle that 'All men by nature desire to know'; the human brain has evolved to ask questions, and to survive by answering them. Science is simply the formal realization of our natural curiosity. Yet no one could fool themselves any longer that, as we stand on the cusp of this new century, we are travelling the simple path of 'progress'. Sure, for several generations now we have strived to balance the pay-off between 'unnatural' mechanization and a pain-free, hunger-free, longer-lasting existence; but now we face a future of interactive and highly personalized information technology, an intrusive but invisible nanotechnology, not to mention a sophisti-cated and powerful biotechnology, that could all conspire together to challenge how we think, what kind of individuals we are, and even whether each of us stays an individual at all.

For The Cynics the implications that this prospect poses, in all its horror and excitement, will be sensationalist hype at best and scaremongering at worst. They won't believe that science will ever be able to produce new types of fundamentally life-transforming technologies, and even if it were, they feel that humans are sufficiently wise and have an inbuilt sanity check to deal with any ethical, cultural or intellectual choices that might ensue. This attitude is not only questionable – in the light of the far more modest precedents that we have witnessed in technology over the last half century – but also chillingly complacent. Can we really afford to assume that humanity will be able to muddle through? And even if we did survive as the unique personalities we are now, in a world bristling with biotech, infotech and nanotech, can we still be sure that such passivity, just letting it all happen, will be the best way to optimize the benefits and reduce any ensuing risks?

Perhaps both Technophiles and Technophobes would agree on one very important issue that sets them aside from The Cynics: we must be proactive and set the agenda for what we want and need from such rapid technical advances; only then shall we, our children and our grandchildren come to have the best life possible. So first we need to evaluate the 21st-century technologies, and then unflinchingly open our minds to all possibilities . . .

2

Lifestyle: What will we see as reality?

Humans were once imprisoned by the dark. As night blacked out the hills and trees our ancestors would have become powerless – watching the world shrink from the bright, wide field of daytime to the shivering compass of a candle. The flickering of flames on a wattle-and-daub wall would have imposed a different reality of long shadows and uncertain shapes; and always, waiting beyond the edge of light, would have been the unknown, unexplained, half-imagined forces. Until merely a century or so ago our ancestors must have lived a life where the arrival of dusk never ceased to make them utterly vulnerable. They would have had a collective mindset of fears and wonders that is now almost impossible to comprehend. But, as we stand at the beginning of the 21st century, the human mind could be on the brink of a makeover even more cataclysmic than that which separates the attitudes of these earlier generations from our own new-millennium view. How will the new technologies transform the ways in which we see and understand the world? More immediately, why should these advances have any more impact than the inventions and discoveries that have already underpinned the ebb and flow of previous civilizations?

Increasingly, it has become the norm to pass people in the streets talking animatedly to no one, single-handedly engrossed in some private crisis: a contemporary illusion of insanity spawned by the ubiquitous cell phone. The haphazard and hectic and intimate press of humanity ebbing and flowing in the shopping centre or on the train, chance smiles and fleeting eye-contact, are a fading reality. Yet imagine a world in which the mobile phone is miniaturized into visual invisibility, insinuated into your clothing, and requiring only a minute amount of power, which can be generated by your own body's reactions. There are no cables or earpiece paraphernalia – your interlocutor might just as well be next to you, but almost in a different dimension. Imagine such a world, where the streets are filled with people lost in invisible, one-sided dialogue all the time, and oblivious to the reality of the physical present around them. The public places of tomorrow, then, might end up not being public anymore, in the way that the park or market or piazza currently imposes spontaneous and intense interaction with others. Instead the high street would become merely a space where many people happen to be, bustling past each other but oblivious to each other as they chattered and laughed and shouted and complained within the dimension of a different reality. Crude caricature though this may be, few would deny that increasing numbers of people are behaving in this way. By offering a route into a parallel world that disregards the messy present perhaps the cell phone is a technical augury of a way of life to come.

The notion of an alternative to the immediacy of the real world may soon extend significantly from public spaces to the most private of private spaces: the home could also end up looking and, most importantly, feeling very different. Homes have always been our ultimate retreat, the place where we are free to walk around naked, speak our minds, smash the crockery, swing from chandeliers – above all to have complete autonomy over the sounds, sights, smells and tastes of the immediate physical environment. Ever since our ancestors had sufficient technology, money and time to choose wallpaper, hang up a photo or shift ornaments on the mantelpiece, we humans have seen our homes as extensions of ourselves. So if, in this century, our minds are set to be transported beyond the press of the moment, we should not be too surprised that our homes may be both a reflection of, and an important influence on, completely novel ways of experiencing life.

Given appropriate sanitation and sufficient resources, the human home has adhered to a pretty much familiar pattern. Many of us still live happily in houses built a century or more ago. For as long as we continue to have bodies clamouring with a range of physiological demands it is hard to imagine not continuing to follow the old arrangement of different rooms for different bodily functions. No amount of high-tech could make it attractive to use the same area, however large or bright, circular or sloping, for real-world cooking, sleeping, washing and lavatorial pursuits. Similarly, the atavistic habits of eating together and defecating in private seem impregnable; the classic Luis Buñuel film *The Phantom of Liberty* had great impact simply because it questioned the age-old norm for dining publicly and voiding bowels and bladder alone. Guests sat around a room, conversing with their hosts, all perched on their own lavatories. Meanwhile, as hunger pangs struck, each individual would whisper their apologies, slip away and seek out the 'smallest room'. Once there, they would bolt the door, take down a tray from the wall and eat swiftly and silently in isolation.

This scenario is as illogical as it is unlikely. There is no immediately obvious reason why we should suddenly change our sense of taboos and socialization ('companion' after all derives from a 'sharing of bread'), and if such habits are to change, it will be because we no longer wish, for new, modern reasons, to socialize rather than because we are answering a call from some deep-seated nook or cranny of the human psyche. On the other hand, our predilection for many small and cosy private rooms over spacious, more communal 'flexible' areas is surely not only a matter of eras and evolution, but of personal taste and particular family requirements.

Just look at the wide diversity today of the kinds of living spaces different people favour. The big change this century, however, may well be that we spend far more time at home; as the real world becomes more dominated by IT, both work and leisure will have the potential to transcend space restrictions. And changes in the way we use space, and what space we use, might sound like science fiction now but could have clear practical advantages in an overcrowded or ecologically compromised planet.

In the distant future, perhaps, our successors will be living in lighter-than-air mega-structures, or in oxygenated sub-ocean habitats, and thereby be able to move vertically as easily as horizontally. Such a scenario would enormously increase the occupancy space on the planet; and, on a more personal level, this type of vertical living would change dramatically how the senses shape the mind, rather as incessant pain from an aching tooth or the misery of outside toilets formed the outlook of bygone generations. Future humans would have a completely different feel about themselves reaching up and down, as well as sideways, in the 3D world. Constant exposure to a certain set of inputs from the senses, like the endless nights of flickering flame of our ancestors, must surely impact on how a mind, any mind, saturated with such sensations, ends up thinking.

Already, a Melbourne-based design company called Crowd has introduced the concept of the Hyperhouse, in which a central electronic spine enables appliances to be moved around and plugged in at different locations, whilst the floor plan and walls are completely flexible. So, kitchens and bathrooms need not be the fixed points in a turning world that they are today; rather, all space can be used differently at different times. This idea of endless change, and of endless possibility, is echoed by designer Fred Blumlein: 'The house of the future will be sort of like having a servant. With an automatic, whiz-bang, digitally controlled environment, you'll make a wish and your wish comes true.'

We are not talking here of clunky gadgets that in a gimmicky old sci-fi way emulate basic human actions, like opening doors. Rather, the encroachment of IT and the pull of the virtual world will mean that, for better or worse, you are hardly ever truly alone. And as the boundaries smudge between yourself and the outside world, between you and your wishes, you yourself might become a vaguer phenomenon.

Moreover, if home-offices are to blend into individual private areas with work-stations, all effectively under one roof, then we can expect a reversal of the ratio of computer-dominated rooms to old-style, non-virtual living rooms. The virtual, computer-run areas of the house would increase until the presence of technology might seem inescapable. It would be a world in which we might still feel the need to preserve certain aspects of our privacy and of our social selves. Future homes might feature a single room left as 'natural' or 'real', the ceiling stripped bare of videoconferencing devices, medical and security scanners, heating, ventilation, air-conditioning sensors and environmental regulators. Just as the opulent home of today might boast a sauna-room that is used for a relaxing communal activity outside the usual routine of daily life, so a 'real room' might have similar appeal – without the luxury or elitist status. After all, it would be hard to have a spending spree when the room was devoted to an *absence* of gadgets and facilities.

In most of the house, however, the IT-dominated environment will permit emptier and more multifunctional spaces: fewer, larger rooms might become the norm, compared with the present-day tendency for many small areas with fixed furniture for fixed functions. Moreover, there is no reason why the boundary between indoor and outdoor living space should not succumb to the same trends, and the distinction between home and garden, already fading, become completely non-existent by 2020. The notion of 'outside' versus 'inside' might cease to have any relevance regarding meteorological caprices: almost sixty years ago the architect Buckminster Fuller drew a fantasy, a dome over New York. It is hardly stretching the imagination too much, therefore, to conceive of large-scale residential areas where the outside is as regulated as the inside.

These large-scale changes might sound very futuristic but temperature-sensitive paint, developed by a company in Shanghai, is already with us: this innovative substance cools your home in summer and warms it in winter. At the same time crystal violet lactone, a thermochromic substance, produces, as its name suggests, a variety of hues at differing temperatures. In winter the house would be a warming red, and in summer a cooling blue. Flexibility in environments, and the speed with which they can adapt to changing needs or circumstances, will be the buzzword, and the yardstick.

In order for us to live in such an endlessly changing environment the next generation of smart heuristic devices will not slavishly carry out preprogrammed operations but will function on an 'if X, then Y' basis. Domestic appliances that are not just automated but also proactive will be the staple of everyday life, and this may even extend to gadgets for leisure: toys, ping-pong sets and even musical instruments. For example, instead of scraping away on a violin, you might prefer to use a recreational simulator whose crafty software could work on your ham-fisted efforts to simulate a Stradivarius sound effect.

So you could be living in an environment where all the domestic, work and leisure devices that nowadays you switch on manually instead burst into operation at your spoken commands. In time this spoken command will no longer seem silly or embarrassing or even downright impractical as computer-controlled smart objects, 'things that think', become part of everyday life. True, it has so far been quicker to open doors with our hands, press buttons with our fingers and turn real keys in real locks. Any intervening electronics have seemed clunky, time-wasting and therefore silly gimmicks rather than true assets. But in the future the difference will lie in the personalization of those objects, and hence in how you interact with them. Future generations will customize the volleys of mechanized voices that will tear through their every waking moment. Such voices, and such machines, may well end up as friends, or rather servants. Presumably it will be child's play, literally, to choose the gender and accent of domestic appliances, even perhaps customizing the voice patterns to sound like those of friends or celebrities. Your alarm call could be Marilyn Monroe or Leonardo di Caprio cajoling you by name to wake up, or more bizarrely, your own voice ... This is not a catalogue of 'wow' phenomena and inventions, but rather the beginning of understanding how we might think about ourselves in a rather different way, and how the boundary between ourselves and the outside world might start to blur.

But remote yet personalized interactions with smart domestic appliances will be relatively trivial in everyday life. Bigger transformations will already be afoot that will influence thought processes – for example in obviating the need to learn another language by using a multi-language translator phone. Similarly, but counter-intuitively, the computer might take the familiar form of a book, at least an IT version. Neil Gershenfeld, in his timely *When Things Start to Think*, catalogues the virtues of books over computers. Books boot instantly, and have a high-contrast/high-resolution display; they are viewable from any angle, in bright or dim light; they offer fast random access to any page, with instant visual and tactile feedback; they are easily annotated with no need for batteries or maintenance; finally, they are robustly packaged. By contrast, the laptop meets none of these specifications.

How could the user-friendly aspects of a book be combined with the efficiency of IT? Reusable printer paper is already possible; it uses e-ink contained in the paper by a micro-encapsulation technique. The next step would be to print wires and transistors on paper to form circuits, so that a radio receiver could be integrated with the paper. This device could be run on solar energy, so that if the 'paper' were left out in daylight to power its circuits, a radio signal could deliver a news update each day: the morning paper of the future. As Gershenfeld has pointed out, the only disadvantage of books is that they convey static information, whilst computers can give changing information. This new system would offer all the benefits of the book, coupled with the dynamism of IT.

Whilst IT might spread into the area of paper publication, the Principal Investigator of MIT Media Lab Research Group Projects, Hiroshi Ishii, is stretching the concept of a computer in another direction. We are all used to dealing with an audio and visual interface – watching the television or using the computer screen. But imagine a world in which we were also used to exploiting the medium of touch for communication. The idea is that as the gap between cyberspace and the physical world is narrowed, and the two become increasingly inter-related, IT will work in more and more areas. As Ishii says: 'We are seeking ways to turn each state of physical matter – not only solid matter but also liquids and gases – within everyday architectural spaces into "interfaces" between people and digital information.' So the world itself would become the interface, with the user detached from reliance upon a computer screen. One important factor of these 'tangible bits' is that the pressure of fingers could form a touch-based communication system, so that the blind and deaf might be able to transmit thoughts. Not only would such a system help those with visual or auditory impairments, but it would also provide a further dimension for everyone in making the cyber-world ever more 'real'. Future 'touch-typing' courses would actually entail learning to communicate by the medium of touch itself!

Think what it might be like to live in the second half of the 21st century, and take a walk around the home. We'll start off in a non-committed space that might be the descendant of the turn-of-the-century living room. As in its 20th-century incarnations, spaces for 'lounging' or the even older activity of 'sitting' when your body has no immediate requests, no specific physiological function will be catered for here. On the contrary, this is the place to come when all internal drives are sated. As you enter the walls start to glow, because banks of light-emitting diodes are linked to monitors that detect your presence. The precise colours vary, as you voice-activate the wall-control system to change from the colour grey, say, to blue. Eventually you may not even have to speak. As sensors in your body detect that you are becoming agitated the walls will automatically turn a soothing cool hue. Perhaps you wish to superimpose a design, and whisper 'summer sky' – whereupon some white fluffy clouds waft around the corners and on to the ceiling. Soon you'll be feeling as calm and fresh as if you were truly out of doors. This sensation is all the more immediate thanks to the air-control system, which is filtering out such pollutants as there still are and adding the desired smells of the moment, such as a sea-air sensation that will automatically accompany the 'summer sky' wrapping around you.

You breathe deeply and start to feel calmer. Should you feel stressed or on the verge of depression, the shifting chemical landscape of the tissues and fluids of your body will trigger sensors that transmit this information to the surfaces of the walls, which accordingly take on a rosy pink hue. In fact, such relatively crude and definitive feedback is not needed for this transformation to occur – even gestures or voice tone (prosody) will be monitored via a scanning laser range-finder system. As you walk around your movements activate 'responsive portraits'; some are of your family and friends shimmering in their frames. They smile in greeting as you pass by, and from wherever you turn to them in the room they are sure to be following you with their eyes. Other pictures are holographic, 3D images, which might utter recorded messages as you look at them. They will, of course, automatically mute when the phone rings, or if you issue a voice-instruction, or, perhaps the most improbable of all, begin to hold a conversation in real time.

But your cyber-reverie is rarely ruptured. You can, after all, trust the computer absolutely with all everyday tasks, such as checking up on your car. The latest engine problem has been remotely detected and corrected by downloading software repairs directly over the internet connection, and you do not really need to oversee such a routine procedure, any more than you have to make any conscious effort to pay. You could, naturally, beam funds into a vending machine, till or bank account by a digital squirt on your mobile. But the issue of automatic payment was settled once you had placed your order. It does not cross your mind that you might have been overcharged.

Everyone now uses smart money, which will transfer itself from your account having first ensured you have paid the lowest rate. If not, the car company, like all businesses these days, is aware that the difference will be refunded. In any event money, at least the physical commodity, no longer exists. Spending patterns are monitored and analysed automatically so that no one can get into too much debt and you are always aware of your immediate financial position.

Untroubled, you move towards the window and smile at the nostalgic memory of the days, which you can just recall, when some people still used curtains or blinds. At a word, the glass darkens with the technology of electrochromism that will, when the world itself darkens outside or on the word of command, convert the transparent glass panes to opaque hues. Then again, you might prefer to instruct that artificial, full-spectrum daylight stream into the room. Another possibility is to distance reality still further and ask the voice-activated window system to display the peak of Mount Everest, a Caribbean beach at sunset or a Cornish lane in springtime.

In all cases, inside the room you are surrounded by light, but not light bulbs. There are no light fittings as such, no special device that is there solely to emit light. Instead, everything is iridescent and provides subtle pools of illumination of different intensities. Light-emitting polymers, as thin as paper and as flexible as fabric, substitute for conventional coverings for the soft furnishings. You sink into a glowing cyber-chair. Sensors in the chair register your particular sitting pattern, comparing the distribution of pressure from all over your body to the sitting profile programmed previously. The voice system in the chair addresses you by name and advises a change in position that will be better for your spine. But you decide instead to voice-activate the zero-gravity system. Your body acquires the semi-foetal position in which astronauts float. So, now you are floating and you can truly start to relax. You wonder aloud what might be on TV and the sharp image of one of the many digital channels appears on the wall opposite you. You command the image to spread out over the whole surface of the wall opposite, knowing that if you swivelled in your chair the image would be transferred, triggered by movement sensors in the chair, to another wall for optimum viewing from that angle.

Yet the term 'TV' is really now a misnomer, with only the name linking the current system to its distant 20th-century predecessor. There are obvious technical differences, like a sharper image on the screen; high-definition television uses about twice as many lines as the standard 625, with a bandwidth some four times larger than the old PAL system. You can hardly imagine what it must have been like to have been the passive victim of the programme scheduler; *that* disappeared within the first decade of this century. Interactive TV goes back as long as anyone can remember. Nowadays everyone takes for granted that all films are available through the TV system, whilst you can access programmes as soon as they are made, so that everyone personalizes their own schedules. TV is no longer part of an external reality, a secondhand real life controlled from the outside; instead it is part of your subjective fantasy world, tailored for you, and you alone.

Not that you need to tailor it much anymore. Your particular preferences are programmed into the system and updated in the light of each schedule you develop, so that increasingly you need not actively deviate from the preselected set of programmes, picked for you by the system; even the adverts cater specifically for your personal predilections. The commercial break has therefore been transformed from the irritating distraction that it used to be into the core method of shopping. After the two-minutes sales pitch you can order the item immediately, by telling the TV to go ahead. Of course, the system already has all the relevant information it needs, from your credit details to your neck measurement.

But tonight you just want to relax. You whisper '3D mode', and immediately the televised personalities and events come down from the wall and surround you. You are immersed and involved – not least because TV is now intensely interactive. You can choose the camera shots you want, and choose the ending to a drama at a spoken command; tonight you request a romance and reject an unhappy ending. The TV system – partial virtual reality, PVR – operates in parallel with the real reality of your living room. The characters in the programme are like ghosts, fighting and arguing and debating right by your coffee table. Yet your coffee table is still there.

Yet even the PVR does not manage to dispel your end-of-day angst, and there is always the possibility that someone else in your household might want to share the same physical space. As it did for the bygone commuters of the turn of the century, the answer lies in a miniaturized personal space – a more comprehensive system that nonetheless has clear similarities with the old walkmen and cell phones. In this case, you can blot out the real world by slipping into the micro-environment of the cyber-chair itself. A hood rises up and over your head, surrounding you with sights and sounds that shut out the immediate room. Now fiction is total: virtual reality at its most personal and intimate.

From the cyber-chair in the living room it is a small step to the cyber-bed in the bedroom. As you yawn and shuffle into this next space, you could remark, if indeed it were still remarkable, that most of the systems that make your environment personalized and inter-active in the living room are duplicated here. The only difference is that there you were more or less vertical, whereas this is where you come when you wish to be more or less horizontal, for more or less all of the time. The specific activity that happens here is sleep. At a further whispered command, you are led into a world that lures you away from consciousness – the soothing experience of a deserted beach, where the waves rhythmically hiss on to the sand, and suck back again, and again. The sun is fading, and you close your eyes . . .

As soon as your body sensors signal to the cyber-bed that you are in the first stages of unconsciousness the beach programme turns off. However, the cyber-bed stays on alert and monitors you throughout the night. If your body temperature becomes too high, then the ambient temperature in the room adjusts accordingly; throughout the night, sensors continue to measure your blood pressure and heart rate, as well as the degree of activity in tossing and turning, and the electrical activity of your brain.

Every morning you can inspect how you fared whilst asleep, and discover how your health is generally, as you peruse your read-outs from several days. Of course, if there is a problem, there will already be a voice-alert for you to consult the medical services. In the case of an emergency situation, you would, needless to say, have been roused in the night. Once people might have regarded the recording and reporting of every glitch in your bodily function as a matter of concern: after all, such information could be made available to any accredited third party, or even a hacker. But now it is all as normal as credit checks were a few decades ago. You are used to living if not in the public eye, as some did in the early 21st century, then at least in the eye of an omnipresent and anonymous third party. In fact, you pause for a moment to think how strange it must have been in the past to have led a completely private existence; no one would have known anything about your personal life, but then again much would have been unknown or perhaps baffling to you yourself, and in general you would have had little control over both the immediate and long-term events unfolding within your body and your mind. In some strange way, somehow, it would have felt as if no one really cared – and indeed it must have felt very isolating and lonely as well as unpredictable. You find this prospect of a private life, buffeted around by happenstance from one moment to the next, actually quite frightening.

In any case, this morning all is well. Your wakening has triggered, via your body sensors, a hot cup of coffee, brewed in the machine at your bedside, a remote descendant of the coy little suburban Teasmade from the second half of the 20th century. This is just one simple example of how the deeply ingrained social and sensual aspects of eating and drinking are now meshing with the new technologies with a variety of outcomes. So what is happening in the kitchen?

All appliances here are for the preparation of food and, irrespective of their actual function, the shared feature is that, of course, they are all 'smart'. For example, your fridge 'knows' when you are low on milk or any other item that you use regularly. In fact, the fridge knows how often you usually use each item and can send orders to the supermarket, via coded identification tags, to restock and charge your account automatically. The barcode on each product is read off, and, as supplies run short, orders are placed and delivered to the door. If you wish to break with your normal habits and order something different, then, as well as the usual and well-established means of internet shopping, you can make a direct visual link to a supermarket and direct a shopper to cruise past each of the shelves on your behalf. You can discuss items with the shopper, who will supply additional information as to the source of the food, best recipes and anything else you might need to know. The shopper is, of course, virtual.

Nowhere are the old and new more intrinsically combined than in the kitchen. Food remains obstinately in its old-fashioned bulk form.

Nutrition pills, so confidently predicted by 20th-century sci-fi comics, exist now, but who would want to eat them? Although technology is able to encapsulate all the nutrition we need in a pill, the enormous pleasure derived from cooking, chewing, tasting and swallowing food ensures that it is still recognizable as such. Food pills are the fall-back for fast refuelling, as junk foods were at the turn of the century. However, ever since the fast-food outlets were sued for masses of damages due to the obesity, diabetes and cardiovascular diseases caused by high fat and salt intake, so crisps, hamburgers and sweets are now as obsolete and as reviled as cigarettes had become by the first decade of the century. So now food pills may well be fast, cheap and convenient to eat but producer and consumer alike are very anxious that they deliver the optimal nutritional package.

Still, food pills are eaten only when you have no time for 'real' food. Although much has changed over the last few decades, the notion of cooking as a rewarding way of spending time has persisted. *How* you cook, on the other hand, has been completely revolutionized. For a start, the physically isolated and extremely clean conditions in which everyone lives have led to a general predisposition to infection and allergies that is far more widespread than in the 20th century. Everyone is now far more sensitive to microbes, toxins and contaminants than they used to be. All the surfaces in your kitchen have been treated so that they are bacteria-repellent. Preparation of food is taken very seriously in terms of hygiene, and you first have to scrub down at the committed hands-only sink in your kitchen.

You think what you might cook up for breakfast. You could use your 'flashbake' oven, which cooks as fast or faster than a microwave, whilst crisping or browning the food in a manner similar to that of a traditional oven by combining microwaves with other heat sources. A screen in the oven-door shows instructions, and an encouraging voice guides you through a recipe; sensors can read the same identification tags on the food that were previously monitored by the fridge. Alternatively, you can request a master-class from a top chef to guide you through the steps of an elaborate recipe. The chef not only explains what you have to do and how to do it but also provides interesting tips and answers any queries you have as you proceed. If you wish to know still more, the maitre will describe the origins and cultural history of the food you are preparing, discuss its nutritional features, as well as suggesting accompanying dishes and wine.

As you finally serve your meal you access a nutrition read-out regarding calories, fat content and all else necessary for you to appreciate how well, or how poorly, this particular meal fits in with your personal daily dietary requirements. This information has been generated at the start of each day from the amassed body data that was collected whilst you slept. Your recommended food intake is estimated using the database of your usual energy requirements, age, weight and usual intake, and now any small shifts in your weight and fat content that may have occurred overnight, as well as additional information that you are about to spend an unusually energetic or lethargic day, will be taken into account.

Occasionally you like to break out of normal behaviour patterns and indulge in a ready-cooked meal not anticipated by the smart fridge. So you order a takeaway. Through your voice-activated central computer system, not only can you see your ideal pizza but you can also sample its taste and smell before you order. Even as far back as 2002 there was the fledgling technology to create a desk-top printer that 'printed' tastes and smells, by means of a cartridge with hundreds of water-based flavours deposited in specific combinations and amounting to over a thousand different smells.

Still, all food, whether home-cooked or takeaway or a mere pill, comes from genetically modified produce. You settle down to your meal – this is one of the few remaining experiences that are 'real', involving an interaction with the atomic, physical world and a direct sensory experience solely within your physical body, and no one else's. As you indulge in chewing, sniffing, tasting and swallowing – you reflect how attitudes to GM foods have changed in a relatively short period of time. The first GM foods contained the actual substance of the source organism, as tomato purée does; more frequently now, however, they consist of purified derivatives that are actually indistinguishable from the non-GM organism, such as lecithins and certain oils and proteins from soya. Since the GM lecithin is chemically identical to its non-GM counterpart, it is hard to see how it presents any additional health risk.

The problem at the turn of the century was the impossibility of guaranteeing the purity of each substance. Nowadays, focused tests are still needed to ensure that alien additional sequences are not absorbed by human tissues or into gut micro-organisms. GM foods have now, for several generations, been part of the culture with no disastrous consequences, so the public is more accepting. Opinion first began to change when people realized that there was simply no alternative way to feed what was then known as 'the developing world', that vast majority of technologically disenfranchised humanity that cast such a shadow over the achievements of the early part of this century; according to UN estimates of the time, 800 million around the globe were undernourished. As well as combating mass starvation, the use of GM foods has reduced the number of children suffering from blindness due to Vitamin A deficiency from 100 million to zero. 400 million women of childbearing age no longer suffer from iron deficiency, which had increased the risk of birth defects. Rice is now routinely engineered to contain beta-carotene, ready for conversion by the body into iron and Vitamin A. And pests had been a problem for crop-growers: for example, 7 per cent of all corn had had to be destroyed. The use of GM crops, engineered to be pest-resistant, eliminated such wastage, as well as providing an attractive alternative to highly toxic pesticides.

However, the biggest incentive for wholesale acceptance of GM foods ended up being, quite simply, personal gain. Interestingly enough, the gain now has nothing to do with health or nutrition. You look out into a world that is no longer composed of the muted mists and shades of previous centuries. Instead, because much of your time is spent processing artificial, heightened and bright colours screened on the windows and walls of the home, even the bright colours of unmodified carrots, spinach or tomatoes would appear dull and unappetizing compared with their iridescent modern counterparts. Now that food has become the main vehicle for that rare and prized phenomenon, direct stimulation of the senses, manufacturers have realized that by genetically modifying the non-essential features that sledgehammer the senses they can make their consumers happy.

For a long time, the taste of each ingredient has been genetically engineered to be much more intense. Any edible substance can take on the flavour of any other. This flexibility is just as well since the production costs of 'real' chocolate, for example, preclude it as a viable product, as do the prohibitive amounts of sugar it contains. The colours of foods, too, are far more vivid and standardized. And not just colour – the physical form is more attractive too. Food also comes in a vastly more varied range of shapes and sizes. Thanks to genetic engineering, as well as precision manipulation of the atoms that constitute matter (nanotechnology), you can now choose to have cuboid vegetables or meat in geometric shapes. So bulk stacking in the fridge is easier, and the fridge can monitor consumption rates more easily, as the scanning tags are all in a uniform position.

Another advantage over 'natural' food is that everything is now bio-engineered to be easier to prepare; for example, grains now come with encapsulated liquid that is released when microwaved. Dripping, messy sauces are also a thing of the past and are now integral, not free-flowing but magnetized so that you can rub them off with a fork or a finger. Moreover, it is almost impossible to obtain food that has not been enriched with vitamins, minerals and other agents such as fish oils that ensure optimum body maintenance. The neutroceuticals industry is flourishing as never before. Potatoes are engineered to combat constipation; cucumbers come in a purple variety, with as much Vitamin A as cantaloupe melons; carrots are maroon in colour because they contain beta-carotene in massive amounts to improve night vision; and salad dressing will lower cholesterol with regular use. Particularly popular is the bespoke neutroceutical option: food is custom-engineered for you, not just to deliver your particular taste preferences but also to cater for your particular health requirements. Your personal medical profile is monitored and fed into a continually updated programme that engineers foods with supplements according to your particular needs. Yet, like everyone else nowadays, you cannot easily draw a line between the genetic engineering employed to produce interesting food that hyper-stimulates your now cyber-jaded senses and that designed to optimize your health by direct intervention at the genetic level, be it for prevention or cure. Then again, everyone now is aware that it is prevention, rather than cure, of disease that is very much the dominant medical strategy.

Gene therapy is now finally in common clinical use, whereby rogue or aberrant genes linked to disease are modified, ideally before they can realize their unwelcome potential. The main genes for cancer have been identified, and there are now various ways in which it is possible to intercept and block tumour growth. As well as manipulating genes there are a variety of alternative ways to manipulate their end-products. The new class of monoclonal medication has proved more powerful for tackling cancer cells than the naturally occurring antibodies of the immune system; in addition, there are the angiogenesis blockers that starve a tumour of its blood supply, as well as ever more vaccines.

Everyone acknowledges now that cancer is triggered by a range of factors, such as tobacco, diet, toxins, radiation and oestrogen, and lifestyles have therefore changed dramatically to reduce these triggers. At the beginning of the 21st century, 70–90 per cent of all cancers were deemed to be related to environment and lifestyle. Now, fear of cancer is really a thing of the past. Even by 2020 there was a complete encyclopaedia of all the genes linked to cancer, that either triggered or prevented the growth of a tumour.

Old age, so rare in nature and in pre-20th-century human history, because of injury and infection, is now the norm. In part old age has been made a comfortable experience by the prevention of diseases, either those inherent in the body, such as cancer, or those caused by external viruses. A central factor in ageing was identified way back in the previous century: the destruction of the fragile membranes of the cells of the body by aggressive molecules known collectively as 'free radicals'. Despite their exotic, anarchic-sounding soubriquet, these small molecules merely have an unpaired electron, which nonetheless makes all the difference. The once innocuous chemical now becomes very reactive, and interactive, with the molecules that make up the structure of cells. These ensuing interactions (a blanket term for which is 'oxidative stress') lead to many unwanted consequences, including the fraying of the protective ends on chromosomes, telomeres, which have been likened to the plastic cap on a shoelace that prevents it unravelling. Once it had been discovered that the famous prototype clone, Dolly the sheep, had shorter telomeres than would be expected for her calendar age – and far more commensurate with the age of the sheep from which she was cloned – people began to be more circumspect about cloning. The debate about how directly shorter telomeres could be linked to physical ageing continued into the 21st century. But though the premature fraying of telomeres was a concern, at the same time a potential retaliatory strategy was discovered: the enzyme telomerase. Telomerase is usually present only in sperm and eggs ('germ cells'), where it keeps telomeres long, and prevents them from deteriorating like all the other 'somatic' cells in the body. This process inspired a new anti-ageing treatment. If telomerase could be arranged to do the same job in the rest of the cells in the body, then they too would have far more efficient, long telomeres. And if chromosomes had long telomeres, then they would not end up sticking together, and the cells that they govern would, in turn, survive for longer; the body that those cells make up would therefore survive more effectively.

Another common therapy nowadays, pioneered at the beginning of the 21st century, involves stem cells. Stem cells are cells that are in an early stage of development, and hence highly adaptable, depending on the environment into which they are placed: they will become neurons if placed among neurons, or cardiac cells if grown with heart tissue. Technology exploiting the versatility of stem cells had already been realized by the turn of the century, whereby new organs and body tissue might be grown to replace defective ones. The only technical drawback then was that unfettered cell division would continue long after transplantation, leading to an increased possibility of tumours in the host tissue. However, stem cells are now engineered to divide only at temperatures hotter than the body (around 40 degrees), so that when introduced into their new and cooler environment of 37 degrees, further proliferation ceases. The ethical issue about working with human material, like the concerns about GM foods, rapidly dissipated once the clear benefit was established.

Another cornerstone of late-21st-century healthcare is nanomedicine: miniature devices patrol your body, giving early warning of possible problems or delivering just the right amount of medication to just the right place. These new therapies, combined with advances in traditional treatments and knowledge of disease as well as healthier lifestyles, all add up to longer life expectancy. A thousand years ago life expectancy was just 25 years, but even by 2002, in Britain, men and women could expect to celebrate their 75th birthdays, and a female born at that time already had a 40 per cent chance of living 150 years. By 2050 there were 2 billion people over 60 years old worldwide, with such 'seniors' making up a third or more of many populations. Much had happened over the previous few decades for such long lifespans to be truly the norm.

Everyone now recognizes that ageing is not a specific disease but a general deterioration of the many processes that sustain body function; of these, the most feared is still a decrease in vitality and often degradation of intellect. Now science holds the promise not just of protracting mere existence but also of actually extending active life. For a long time now people have been aware of the need to view the mental abilities of older people in their own terms, rather than in comparison with immature, growing brains.

Young people show a 'fluid' intelligence; as its name suggests, the primary ability here is one of ready adaptation, to learn in a quick and agile fashion. Yet as we age, as we all know, this skill seems to decline. If you compare the speed with which a young and old person learn a task, the elderly, inevitably, will be less impressive. But then there is a second type of mental prowess, 'crystalline' intelligence, whereby past experiences are used to assess and interpret the current situation. Needless to say, older people outstrip younger ones in tasks requiring experience and prior knowledge. Perhaps that is one of the reasons, aside from depletion of a ready supply of appropriate brain chemicals and frayed chromosomes, why 'fluid' intelligence is harder for the older brain. There could be a neurological sales resistance to accepting any new process or fact without it first being filtered through preexisting associations and values. However, the big problem nowadays is that there seems little need for crystalline intelligence; facts no longer need to be learnt, and there are no new, unforeseen events that require 'wisdom' to evaluate them . . . The brain can be kept agile by stimulation with interactive IT; and older people as a consequence have kept their fluid intelligence for longer.

But as you stare at your steaming food your immediate concern is how hungry you are. Even though your particular genetic read-out does not place you at any specific risk of a specific cancer, you are obviously keen to ensure that you live as long as possible. Over a century ago now, the effects of dietary restriction on longevity in rats were reported by two Yale nutritionists, Thomas Osborne and Lafayette B. Mendel: they found that rats eating only 50–60 per cent of that consumed without restriction by a comparable group of rats lived significantly longer. Subsequent experiments in more modern times confirmed these findings, and indeed went on to show that the effect was not the result of the elimination of one particular toxic substance, nor was it a simple slowing down of development since the effect still worked in mature animals. Further careful investigation revealed that the abstemious rats were not living longer due to any decrease in body fat, nor due to a decreased metabolism. In fact, the beneficial effect seemed to be generally protective rather than a slowing of senescence.

Although this now well-established phenomenon has only ever been shown formally in experiments with rodents, anecdotal evidence had for some time suggested that the same process occurred in humans. The first such evidence was the Japanese island of Okinawa, where there had been significantly more 100-year-olds than elsewhere in Japan, whilst deaths from heart disease and cancer had been two-thirds that of the Japanese mainland. Interestingly enough, average food intake was 20 per cent less than the national average. The diet had always been the same, just less of it. So why does eating less prolong lifespan?

For several decades now, since the start of the 21st century, scientists have suggested that the cause of this remarkable finding may be that, when the body takes in less food, key enzymes (catalase and superoxide dismutase) increase by some three to four times normal levels; these enzymes neutralize those damaging molecular anarchists, the free radicals that can otherwise attack DNA. But whilst the good news is that reduced food intake means that there are fewer free radicals to do damage, the bad news is that this reduction in deterioration and ageing seems to be accompanied by a loss of interest, in rats at least, in mating. You ponder for a moment whether celibacy is a price worth paying to live longer. You convince yourself that, after all, lifespans can now be prolonged as much as possible by other strategies, such as the use of genetic encyclopaedias coupled with new early-warning diagnostic tests and optimized lifestyle; accordingly, you spoon out more food and eat as many calories as you can . . .

Once the meal is over, the smart garbage system will sort the waste into organic, non-organic, recyclable and so on. It will compact the bulk and eliminate odour. A similar level of hygiene, and a similar level of the individual personal monitoring that occurs in the bedroom, operates in the bathroom. Your scales give a read-out of fat content and weight, as they have been doing for decades; but now you can also check your heart rate, blood pressure, cholesterol levels and immune system status. As you look in the mirror and yawn smart sensors there detect any problems with pupil dilation, tongue surface, teeth and so on. Michael Dertouzoa, Head of MIT Lab for Computer Science, predicted way back, with great prescience, that sensors in the wash basin would be able to detect minor traces of blood from your gums, activating the voice system: 'At the rate you are going, there is a 50–50 chance that you will have a periodontal incident in 12 to 15 months and a loss of half of your teeth by the time you are 55.'

As you evacuate your bladder and bowels your identity is confirmed by buttock prints. Sensors in the lavatory bowl then proceed to monitor your early-morning urine for any signs of diabetes. Your faeces are also screened as they are flushed away, for any indication of bowel or colon cancer, or other defecatory problems. Meanwhile the smart system within the lavatory will in any case issue, by voice, advice on how to modify your diet if your excrement is showing an inappropriate biochemical profile.

These days everyone expects their daily urine test to be sensitive enough to detect tiny colonies of cancer cells; if it does, such early diagnosis, combined with the advantages of genetic profiling, ensures that treatments are as powerful and effective as possible. The process of pharmacogenomics is familiar to everyone; your personal biochip, with a read-out of your individual genetic profile, enables the tailor-made prescription of a drug. Side effects are minimized as the appropriate treatment for you personally is calculated by powerful software that takes into account your history and other medication, as well as genetic risk.

Of course, unlike the unenlightened public of the early 21st century you are familiar with the notion of risk, and indeed of living in a society where everyone is at risk from something. But, by the same token, you do not take risk so personally, since you do not see yourself as an isolated entity facing up to a highly precarious future. Instead you are so plugged in to the collective knowledge base, and so alert to the variety of contingencies that will click into place as soon as there are any early warning signs of potential disease, that you do not see disease itself as a looming threat, nor as a milestone in the course of your life that you, as an isolated individual, will have to deal with proactively. Everything will be looked after automatically by the systems: you do not have to worry.

Given the massive improvements this century that have come with transplant surgery, the growing of organs, genetic engineering, stress reduction, preventative medicine, better diagnostic devices and pharmacogenomics, it is perhaps unsurprising that drugs are no longer primarily used to cure disease, but rather to fine-tune lifestyle. There are now impressive sounding products available that claim to improve sex and IQ, others that can cause weight loss whilst eating to excess, and still more that claim to combat shyness or moodiness or offer 'cures' for baldness, alcoholism, obsessive compulsive disorder . . .

Enthusiasm for these drugs, however, is not as great as when they first appeared. One reason is that they do not seem after all to target the specific problem, and inevitably cause side effects of one sort or another. Secondly, and perhaps more significantly, such 'cures' are no longer needed so desperately. After all, many of your features were determined prior to your birth, when you were genetically screened for 'defects' such as baldness; indeed, all males now have a full head of hair, just as most people in the developed world at the turn of the century had an expectation, unlike those in the early 20th century, of a full set of teeth for life. Bald heads, like the once-prized prospect of false teeth as a coming-of-age present, are curios of past eras. Meanwhile, more elusive mental traits, which do not have a one-to-one relation with a single gene, can nonetheless be effectively manipulated by subtle organization of your immediate environment.

Living in an interactive world, where much of reality is virtual, every sensation you experience every waking moment can be controlled with a precision far more sophisticated and sensitive than potentially hazardous substances indiscriminately marinating the complex biochemical mechanisms of brain and body. The mind can be more powerfully bent with clever software than by the sledgehammer of a substance that changes wholesale the delicate and balanced chemical cascades within the brain. Precision IT has overtaken pharmacology for creating altered states, and indeed determining what altered states we wish to experience.

In fact, IT has not only ousted drugs as a remedy but has also rendered most doctors obsolete. Although the medical profession as such still exists, many feel it is a dying art, akin to blacksmithing at the dawn of the 20th century. It all began with the information overload for doctors that ensued directly from the new technologies, treatments and diseases and burgeoned over the next few decades. By the early 2000s the medical information needed by a doctor in day-to-day practice was turning over every five years, and as this turnover became faster and faster an increasingly common sight in surgeries was a patient clutching sheaves of downloaded information on their particular illness – the cyber-world was already offering a more accurate and faster alternative for both diagnosis and therapy. Nowadays, you need only to access the home medical box software on your bathroom wall for officially regulated in-depth explanations of any disease or symptom that concerns you. Then there are the electronic discussion groups and self-help groups for rare diseases and, finally, hospitals as networked environments. Often even emergencies are treated at home, or in smaller satellite hospitals, since a network line directly links paramedics to experts.

But by this time, any thoughts of illness have dissipated; you have a clean bill of health from the monitoring systems and you are already thinking of what to wear. One company, I-Wear, set the trend long ago by endeavouring to plan clothes for the future. Traditionally, they reasoned, garments had served three functions. First that of barrier – keeping warm, decent and protected. Their second function was communication – even in medieval times the colour of clothes, not to mention their condition and fabric, denoted the wearer's position within the feudal hierarchy (peasants were actually not allowed to wear bright colours). A few centuries later the old school tie and the very wearing of a tie itself were obvious examples of messages sent via our clothes. As society became more tolerant, clothes in the 20th century came to express attitude, moods and affiliations in a myriad of ways. Meanwhile, the third function that had always been important was the organizational aspect of clothes – be it a belt for a dagger or zipper pockets and money belts.

But, more recently, you are finding that you need only a few clothes since they are all made of smart polymers and are therefore alterable. The all-important purchase is the latest software, or 'softwear', to transform the material you have into the desired style, texture and colour. Now you can transform any item of clothing to accommodate your mood, weight, fashion and the occasion. In fact, you can feed in all these requirements and a recommended garment will flash up on the screen, to be downloaded in three dimensions as soon as you voice-activate, 'OK'.

Nowadays, underneath your clothing, body sensors continuously gauge your mood according to your pulse rate and degree of sweat. Such information is important as it will then determine how you receive information at any particular moment from your personalized IT systems. We know, for example, that if people are stressed, information received quickly helps them relax; on the other hand, if they are already relaxed, information transmitted quickly can be highly stressful. All this ongoing feedback about the state of your mind and body is still accomplished by successive generations of e-broidery, initially developed way back in 2002 as conducting threads in fabric; the traces, then as now, carry data and power.

Much of what your predecessors used to carry around, such as mobile phones and reference material, is now incorporated into clothes. You merely interrogate your watch, say, about your location or best route to follow, and receive an oral response. Keys, similarly, are obsolete since iris scanners and voice-recognition systems open the relevant doors. With a spoken word, a thumbprint or a glance you can make transactions that no forger or thief could ever emulate. And all the while your health continues to be monitored throughout the day. The heart monitor in your T-shirt gives a constant read-out. The fabric also inhibits odour and releases perfumes appropriate for different contexts, such as seduction, playing with children, being with friends or working. Moreover, the fabric of all your clothes interacts to counteract hormone or blood-sugar imbalances with impregnated drugs, released as soon as the sensors pick up a deficiency.

As your body temperature and moisture fluctuate so will the feedback from your clothes. This feedback influences anything you choose, from the temperature of the room to the colour of the clothes themselves, which can change in an instant. Your shoes, in particular, are no longer mere protection for the feet but a source of energy; the human body generates, from one moment to the next, some 80 watts of useable energy, of which 1 watt comes from the feet alone. Transducers in the soles of the shoes convert the energy generated on impact with each step, to be recycled when needed to aid tired muscles. If you prefer, the energy can be used to power the 'invisible' computers embedded in your clothes and jewellery. Neil Gershenfeld at the MIT Media Lab long ago envisaged that it would be possible to send an electrically coded CV, stored in the shoes, up into the hands during a greeting, a prediction that turned out to be correct. Because the sweaty palms conduct electricity the electronic CV can be exchanged with a handshake, the modern equivalent of the tradition of exchanging business cards.

Unlike the inevitably shabby and baggy clothes that sooner or later resulted from a normal turn-of-the-century working day, your clothes always look pristine. They are all now made from wrinkle-free fabric, and self-clean as you wear them. The reason that the kitchen is exclusively for the preparation of food is that laundry is truly a relic of the past; bacteria impregnated into every single fibre of fabric flourish and breed by feeding on the dirt, thereby creating self-cleaning clothes. And unlike the clothes of the past, yours now fit you perfectly; when you cyber-shop from your home you are able to indulge in virtual fittings, having fed in your precise body measurements.

The communication function of clothes is now automatic and leaves nothing to chance. Much of your time is spent in virtual interaction from your home. When you have a business encounter it is therefore so much more practical to turn to a virtual clothing package, and transmit your image in virtual clothes added as part of the software. In this way you know that you are wearing the most culturally acceptable attire for the occasion. In turn, the cultural requirements and expectations of any encounter are contained within an appropriate program, developed by anthropologists and psychologists, so that the age, gender and nationality of those you are meeting, together with the location, time of day and season, as well as the desired outcome and the message that you wish to convey, could all be factors in determining your virtual outfit. Some recent, highly sophisticated software now offers the opportunity for you to change your gender, age or nationality so that the virtual you would fit into the meeting perfectly, and stand the best chance of the perfect outcome.

Of course, there is now a dreary predictability to how the participants of such meetings are actually going to look. After all, everyone has access to similar software, so that a meeting norm has evolved where everyone is a middle-aged Caucasian male in a grey suit with an educated accent and vocabulary. Hence you are having to come to terms with the ultimate annihilation of the individual, at least for business purposes. As you see the situation at the moment, this latest system is simply a way of masking any individual factors that would detract from the business itself. The advantage is, obviously, the ultimate in political correctness, paradoxically achieved by exploiting rather than denying human bigotry.

But increasingly you find yourself musing over what 'you' actually are, or indeed what the concept actually encompasses. All achievements, feedback and interaction are increasingly accomplished via a virtual version of the self, sanitized in terms of race, age, gender and background, so what is really left of the real you? Yet no one seems to be raising an alarm, nor even posing the question of what constitutes individuality. Moreover, such standardizing scenarios are now seeping beyond business meetings to social encounters too, virtual dinner parties with virtual guests, for example. Soon your children or grand-children could be living an entirely virtual life with a plethora of multiple cyber-personalities.

No one sees themselves as constrained anymore, either by space or indeed by time. Science is shrivelling global space to a screen in front of you so that there are no boundaries to going where you wish, at least in cyberspace, nor indeed preventing you from going back to any particular moment in time. Since everything has been recorded in real time as you live your life, a flashback complete with all sights, sounds and colours, downloaded to PVR or absolute VR, enables you to relive any time you choose, with the same degree of realism – cyber-realism – with which you live most of your everyday life in any case. The past and present are therefore now impossible to distinguish; the passage of time seems a meaningless notion.

In this way, you no longer have such a clear storyline to your life, compared to previous generations who had to tax their highly revisionist and imperfect, fleeting and fragile memories, feebly aided by photo albums and home videos. But it need not be your own life that you revisit: just as people in the past would watch old movies to escape nostalgically into earlier eras, so now you access individualized software that immerses you completely in another time instantaneously.

If you desire a particularly strong link with the past, as though it were the present, you can even continue an email exchange with someone who has died, because ingenious software is able to extrapolate previous behaviours to mimic the type of reactions and areas of interest the deceased would have had. But things do not even stop there. Once it was accepted that software would be good enough to simulate the reactions of a deceased relation, it was only a small step to make them fictional from the outset. Even your family members can be virtual: they can be of whatever age, gender, sexual orientation, and in whatever number you choose. They ask you if you have had a hard day at work, join you for dinner and even, by means of haptic sensor technology, give you a hug or a kiss.

Even by the turn of the century the internet had reinvented flirting. Email flirting is still growing in popularity because it emphasizes fun rather than a serious relationship – and fun is now the order of the day. After all, there is nothing else to do really except feel the thrill of your pulse changing or your heart rate increasing. One primitive prototype for virtual flirting was a text-messaging service, *Mamjam*, that could send messages from bars and clubs between individuals who had never met each other, but whom the service connected according to their proximity. And now cyber-courtship continues to flourish. Using your embedded and invisible mobile, you fill out your profile and post it on a virtual bulletin board. This profile is designed to fit the tiny screens of mobiles and has spawned a new virtual society with its own words, rules and risks. When this virtual lifestyle first started, 10,000 matchmaking sites sprang up immediately in Japan alone; one site boasted 600,000 members with 3,000 hits every five minutes. Now, of course, numbers have escalated everywhere.

Back in the real world life continues: people continue to embrace real life, if only as a novelty. It has become a kind of hobby, a little like camping without electricity or running water used to be way back in the 20th century. First you have to enter the 'natural room' and take off all IT-embedded and smart clothing. You then sit in a solid, non-smart chair that will never change its function, form or colour, and unless you have made a prior arrangement, you have to wait until another person also feels the 'natural' urge. The next step is to have a face-to-face dialogue, in real time. Your generation finds this activity very frustrating. After all, it is impossible to access information for reference quickly, and you have needed to memorize very little for all other aspects of daily life. What is there to talk about using just your own isolated brain? And even if you did have a way of harnessing whatever facts you might need, what would be the point of this random, slow, capricious interchange with another person? What could they tell you and why would you want to know it? It's all too lonely, too slow, and takes far too much proactive effort. Yet your grandparents still seem to enjoy this primitive activity, at least some of the time . . .

By comparison, however, the more modern ways of passing time are much more to your taste. Thanks to the security systems, wall-mounted iris-scanners checking out all humans moving around your home, you can pass the time studying exactly where every friend, family member or guest happens to be – just as they can spy on you. Similarly, at the end of the day, you can peruse a house log of where and with whom your partner or children spent their time whilst you were away – just as they can for you. Of course, you could turn the system off, but that would immediately invite the question of why you did so. In any case, it's so much part of your life that you are utterly oblivious to any potential intrusion.

The Big Brother of *Nineteen Eighty-Four*, or indeed of the epony-mous TV hit of 2000, certainly did set the scene for what was to come . . . But now there is a difference. The citizens in George Orwell's famous novel, and the volunteers in the TV series half a century later, were the helpless subjects of scrutiny: in each case, they were a cohesive group watched by a distinct, outside force. By contrast, surveillance in your home is interactive: everyone is spying on everyone else. You naturally assume that everyone around you knows everything about how you have spent your time. Indeed, the viewing of daily-life logs, or indeed the real-time watching of friends and family in other rooms, has started to fill up large parts of recreational lives.

Some have even taken things to an absurd limit, and have simply ended up watching each other watching each other. Perhaps someone from the opening decade of the 21st century would have thought that silly, but then did they ever ask themselves what the reaction of a Jarrow housewife in the 1930s would have been if someone had predicted the Big Brother TV series and its popularity? Imagine a family living in a back-to-back slum: the children have no shoes to wear, and brown paper sewn into their clothes is the only form of thermal protection against the cold; there is never enough coal, and always the drip feed of anxiety about not having enough food. Imagine telling such a family that within a few decades their descendants, glutted with far too many calories, would sprawl in front of a flickering screen obsessed with watching ultra-ordinary people live out an every-day life doing nothing. They would have thought that as daft as those living at the turn of the 21st century might have thought you . . .

Attitudes change. House surveillance has surely made modern society far more exhibitionist than it used to be. The idea of turning the system off from time to time because it would have made your predecessors feel uncomfortable is ridiculous; you are not aware of any alternative, and hence of any alternative way of feeling. You are most at home networked into the large, passive collective and therefore do not resent being scrutinized by others. It's more as though they were part of you in any case – a kind of *collective self*.

Whilst our own personalities, here in the second half of the 21st century, have become fluid and uncertain, cyber-personalities – robots – have become an integral part of the domestic scene. HAL in the film *2001* was surely an obvious prototype, in terms of two-way voice communication between computers and people, irrespective of whether he 'felt' anything. The speed and convenience of interaction that resulted, coupled with the increased power of computers, soon led to a new central factor in everyone's lives.

In a trice, everyone was vowing that they couldn't be without their virtual butler, just as in the old days people 'couldn't be without' email or a mobile phone. Nowadays this ubiquitous assistant anticipates your every need, knows everything about you and carries out any instruction immediately and without complaint twenty-four hours a day. As you wake in the morning you see him portrayed on the flat screen facing your bed. You have selected a male persona, since you tend to be a bit of a traditionalist, though many prefer their favourite stars or even ancient icons, such as Britney Spears.

Your butler Douglas is replete in white tie and tails; he reminds you that it is your birthday, but that you have quite a challenging day ahead. He runs through what you have to do, occasionally giving you more information as you query who exactly someone is or why a meeting was arranged in the first place. Douglas is able to interpret your body language and facial expressions; without you speaking he tells you that you seem a bit sulky and enquires whether he should access a new cyber-friend, or arrange for your favourite meal.

Douglas is the latest generation in a long line of less sophisticated forerunners starting with 'Dwain', developed as far back as the 2000s. Of course, no one needs a talking head on the screen for the computer to run your life, but research showed early on, not surprisingly, that people felt more at home with a human-like character. Dwain happened to be a 40-something male persona, but even back then he could just as easily have been a 1950s secretary with stilettos and red lipstick, or had the face of an old friend or dead relative. In any case, you are able to change faces and personas more easily than the old fonts on ancient word-processing software, simply with a spoken word of command.

Perhaps not surprisingly, however, most people stick with just the one character. After all, way back, we all tended to personalize even our cars; although you know that Douglas is not conscious, that he is only virtual, you and most others of your generation still have a basic need to feel you are among sentient beings who really do care. Douglas could, of course, double as a nanny, a stockbroker, a teacher or even a personal trainer, so long as he has access to the output from your body sensors. But just as we used to have a repertoire of players in our lives, from spouse and children to friends and colleagues, you do not stop at one cyber-person, the butler; you also have a constellation of cyber-friends – some people even have cyber-children – all with different appearances and different predispositions to respond to. Since you have real conversation in real time with your cyber-system, it is surely natural to have certain 'friends' that you turn to when you are in certain moods, just as people used to with real friends in the 20th century.

But casual conversations are not as central to your, literally, selfish life as Douglas, who is not only in charge of your home but in charge of you. Douglas is effectively an extension of you, or more accurately of your thoughts and desires and needs – your mind. No need therefore for him to be in more than two dimensions; since there are sensors on your body, and indeed on the furniture, appliances and walls in your home, all the necessary information can feed readily into the central unit. But you still need to move things around in the real, physical world. The only solution is robots.

Far from the creaky tin-men of 20th-century imagination, robots of today have the precision of surgeons, and can be designed to suit the job in hand. Hod Lipson and Jordan Pollack of Brandeis University were ahead of their time, when they envisaged robots made of thermo-plastic that could be easily assembled into three-dimensional structures from a computer screen design, then melted down and recycled for the next job. Eventually the robot ended up doing this under command from a program dictated, in turn, by virtual butlers such as Douglas. But then it didn't stop at robots. Surely, the reasoning ran, if robots can change so rapidly in their shape, then humble furniture can too. Someone turns up as a physical presence to join you in eating, and so you quickly convert a kitchen table into an additional chair.

The end result is that you are used to living now in a world where mechanical devices move around, change shape and replicate under the control of other mechanical devices that can monitor your inner states and anticipate your every secret wish. For example, a personal fabricator is able to generate any type of book you like; say, for left-hand use or with large type, or whatever your specification demands. But the biggest lifestyle change is that you can make devices in three dimensions too. PEMS – printed electro-mechanical systems – provide an effective blueprint for you to assemble any gadget, furni-ture or machine from raw materials ordered on the net. Once you have finished with the object the same system will dismantle it and break it back down into the original raw materials. In this way the old worries about conspicuous wastage and non-biodegradable refuse have truly become a fading folk memory. Yet because nothing around you has any permanency any longer, you view the world very differently from your predecessors. Since everything can and does change you tend to be far more constrained and excited by the actual moment, taking whatever is around you at face value, belonging only to now. The only real reality is the sentient being you call yourself, feeling sensations at that moment.

A further factor in the erosion of enduring individuality is that you no longer have clear and consistent relationships. Children are now expected and encouraged to experiment with both gender roles. The idea is that an individual, with the help of appropriate software and a virtual persona, alternates gender and explores different yet distinct sexual orientations at different times. Hetero-, bi- and homosexual boundaries have long been obsolete, as has the old concept of the family.

Irrespective of marital status, sexual orientation and gender, it was basic personal interactions that defined the family unit. The defining feature was not that a group of people were sharing a roof, as students or a lodger may have done, but that they were actually doing things together for most of the non-working time, be it cuddling, arguing, eating, singing, bringing up children or watching TV. But we now know that TV was a technological molehill when compared to the heady scientific heights of this third millennium. The convincing and attractive cyber-world has played havoc with our sense of space and time. Traditional real relationships in real time atrophied. And the family unit as it used to be, even in its most liberal form, slowly vanished, just as the medieval feudal system, once the bedrock of social organization, faded in the face of new technologies and progress.

Some opposed this trend, even into the first half of the 21st century. After all, it was argued, human beings are social creatures, inter-dependent economically and with sexual and emotional needs to per-petuate the species. The family gave a sense of identity, a feeling of belonging to someone. It was a concept that underpinned most litera-ture – the inter-relations of generations and siblings and their effect on us. There may well have been smaller families towards the end, single parents and complex inter-relations with step-children and step-parents. But then, traditionally, families were always cloaked in secrecy; that was the whole point. You could behave within your immediate relations in a way that you would not countenance in public. But now you do not need to depend on anyone else emotionally, or even economically; your sense of identity is now an enlarged collective one, and one that is essentially public – thereby removing the final advantage of the family: privacy.

The end of privacy was further hastened in 2025 with the introduction of cyber-spheres that monitored and documented every moment of each person's daily life; the idea was that a single network of phone, computer and TV could record every email message, phone call, calendar entry and internet bookmark. All data is incorporated into a cyber-stream of the minutiae of each hour, an electronic life story. So your only enduring stability – your framework for making sense of your existence – comes from a kind of electronic storyline of all your thoughts and activities. Needless to say, the patterns in your life are constantly being analysed to see trends across the whole of society and also to predict what you as an individual will do . . . But you do not have any worries on this score, as you have never known the life that your grandparents used to talk about, when individuals had a kind of isolation from everyone else. The term 'privacy' is an arcane one that very old people use occasionally; no one can really explain exactly what it used to mean. Some say it's the opposite of 'public', but then they say that 'everything nowadays' is public, so you are still baffled trying to work out what the opposite, privacy, could ever have meant.

Central to your reality are your own experiences in a turning world – experiences that can be caught for ever, and indeed touched by others – and central to your comfort, and the backdrop for these shifting sensations, is your home. This home is not merely a place bristling with silly gadgets but a base camp around which images, sounds, textures and smells, and above all information, assault you every second. With the advent of a convincing cyber-world, you work from home and socialize remotely. But your work and leisure are another story, one which we shall explore in due course.

Where do we stand then, as we contemplate the lifestyle that could await us? There could be three immediately obvious consequences for the way we think and act. First, if everything about us, from our buttock prints to our faeces to our movements around a house, is recorded, together with our predilections, mode of talking and types of (cyber-)friends, then surely such information would be valuable. The Big Brother of *Nineteen Eighty-Four* starts to seem very plausible. We would no longer have private thoughts; rather, we would effectively be part of a larger network, a mere node in a thinking, conscious system that goes way beyond an individual mind.

The ebb and flow of this mind will certainly have a persistent effect upon us, and in this way not simply IT but also the presence of humanity within that IT structure will be a crucial factor upon our existence. The pull of the different types of tribes, clusters and personalities themselves within the cyber-society will determine both your 'individual' personality, and society itself. At times, the effects of different communities upon that personality will be either negligible or desirable. But in the case of cults or fundamentalist movements the pull from one direction may be all-consuming.

This question of social interaction leads to the second issue. Will those who live a century from now be socially inept, by the standards of today? If virtual friends replace flesh-and-blood ones, we shall not need to learn social skills, nor think about the unwanted and unpredictable reactions of others. So within this collective consciousness there need be no interaction, no action or response but rather, should we choose it, a passivity in which we are shielded from any disagreement or disharmony. Able to access any information we wish, and capable of choosing from a variety of cyber-companions, what would be the worth in seeking out real-life human individuals? If you were to find them, why would they be interested, or interesting? They too would be busy talking with their cyber-friends or their butler, or watching their favourite film, with, of course, their favourite ending.

If, and it is a big if, you had had experiences that no one else had had, or had access to information unknown to others, then you might have more insight than others. But why should anyone want to listen to your original idea? What would someone else do with your theories of the human psyche? Such knowledge would be of little worth to anyone who does not need to interact with, or understand, unpredictable fellow humans as part of successfully living out their daily life.

Thirdly, like our ancestors staring into the flickering flames at night, our descendants will obviously see reality 'out there' in a very different way. The most immediate step in this transition is the blurring of the interface between artificial systems and humans. As machines evolve from inert lumps of matter into dazzlingly intelligent systems with which we form relationships, and as humans acquire increasingly invasive cyber-prostheses, will the great distinction between silicon and carbon systems still remain valid?

3

Robots: How will we think of our bodies?

Machines will probably surpass overall human intellectual capability by 2020, and have an emotional feel just like people. At some point they will develop genuine self-awareness and consciousness, and we will have to negotiate their rights. By the end of the 21st century, they will have far superior intelligence to people, but probably have more attractive personalities, so relating to machines will be more pleasant than dealing with other humans.

Futurologist Ian Pearson is not the first to predict a future in which, finally, we humans will be sharing the planet with a species that will outshine us, not just in brute mental prowess but also as broad-minded deep-thinking citizens. As long ago as the 1980s Marvin Minsky was predicting the rise of super-intelligent computers 'within five years'; so far, no silicon system has lived up to this estimate. But just how likely is it in the 21st century?

Almost a certainty. Or so it would seem, if you reflect for a moment on how we are already becoming increasingly absorbed by the cyber-world. Consider that familiar fixture of a large number of households today: the glassy-eyed, monosyllabic adolescent in deep dialogue with their screen and keyboard. They are living in a different world, where the inhabitants spend long hours each day surfing the net or sending text messages or playing computer games. The flickering, beeping, flat world on the screen has become as pervasive and real as the pulse and press of the world around them, perhaps even more so. And there is no reason why this trend should abate. The lives of future generations look set to revolve less around face-to-face relationships with each other than around relationships conducted via the medium of the computer, or even with the machine itself as the direct recipient of their attentions.

Although the computer offers a reality of sorts, until now the cyberworld has always had its limits, its boxed boundary easy to encompass with a sweep of the eye. We would need a three-dimensional, all-pervading environment to completely seduce us away from *real* reality. Computers may dominate our lives, but we know where the division lies between screen and 'out there'. Robots, on the other hand – moving about as they do in three dimensions – might have the potential to seduce us away from the real world, were they not still the clunky heroes of heavy-handed science fiction.

Films featuring robots, even Steven Spielberg's recent offering *AI*, usually start with the idea that less than benign scientists are trying to generate human intelligence in a robotic or computerized guise. This scenario is compelling, as humanoid robots most obviously generate a good storyline. In particular, most of us are intrigued by the idea of beings alarmingly cleverer than us but at the same time flawed by a fatal human-like disposition towards world domination. Even the brilliant physicist Stephen Hawking is not immune to this kind of musing. He warns that humans must change their DNA, or else: 'The danger is real that this computer intelligence will develop and take over the world. We must develop as quickly as possible technologies that make possible a direct connection between brain and computer, so that artificial brains contribute to human intelligence rather than opposing it.'

The big worry is clearly that computers and ultimately robots are about to overtake us, although it is not exactly obvious what they are really about to do. Although the development of robots has lagged decades behind that of computers, the new generation of robots is about to get far more serious. At last, the robots of the future will have cast off their sci-fi heritage, having little in common with their predecessors – those tin-men of Hollywood finally consigned to nostalgic 20th-century memory. But why has it taken so long for robots to be anything other than metaphorically hard-hearted, cold-blooded and usually bent on wiping out humans?

Our fascination for the eternal polarization of good and bad, for fairy tales of intrinsically evil machines, is not a good enough excuse for explaining the surprisingly slow technological development of robots in real life. Steve Grand, of the company Cyberlife, has a much more persuasive list of reasons. He suggests that the difficulty may have started as far back as the middle of the 20th century, when visionaries such as Alan Turing were carrying out pioneering work on artificial intelligence. Turing wanted a challenge for very primitive computers that had no sensory device other than a paper-tape reader. Chess seemed a good standard problem as a starting point for such machines. But even Turing himself, it is worth noting, recognized that competence at the game was a poor indicator of intelligence, and that it was important to recognize what he termed 'situatedness', responses to a real, complex environment.

Another reason why robots have not really impressed anyone as yet could be, claims Grand, to do with testosterone – more specifically, the fact that it has been mainly men who have worked on robotic projects. Men, so the argument runs, favour a 'top down' form of control, as ascribed traditionally to the Hollywood robots, rather than interaction with the outside world in a way that would allow an intelligence to evolve 'bottom up'. Until now most of these hormonally challenged individuals have made the erroneous assumption that the brain 'does things' to data; the alternative, and one that is far closer to the physiology of our ever-adaptive human brains, is a two-way street whereby incoming information changes the brain, online, and *at the same time* alters whatever particular and personal evaluations and interpretations that brain had formulated. In effect, the robot should be permanently reprogramming itself in the light of experience. A further difficulty, arguably also arising from the notorious male inability to multitask, is that most earlier 'intelligent' machines have processed information in series, i.e. sequentially, rather than per-forming several functions simultaneously, in parallel. Leaving aside the issue of whether AI (Artificial Intelligence) researchers have not delivered because they are, according to Grand, 'blinkered, domineer-ing, chess-playing (male) nerds', we clearly need to define what exactly we want from our robots in the future.

The goals of true AI have now shifted from creating a 'disembodied, rational intelligence'. 'Intelligence' itself is a very loaded term. A central problem is that it is always defined operationally – whether scoring highly in IQ tests or surviving and thriving on the primeval savannah. Yet when we speak of intelligence, we are usually uncomfortably acknowledging an additional feature, one that actually harks back to the Latin roots for 'understanding'. There is a queasy sense that operational tests and definitions are not really getting to the essence of intelligence – that inner, subjective ability, which is so much harder to test and measure, where we understand something at the visceral level, and do not just give learnt, automated responses. The less emphasis on 'automated' and the more we see of 'learnt' behaviour in robots the nearer they may come to eventually being labelled as 'intelligent'.

The robots of the future will be far more interactive and prepared to learn. For example, scientists at Glasgow University have now developed a 6ft 2in robot known as 'DB' – Dynamic Brain. DB has as many joints as a human, powered by hydraulics. Moreover, a human and DB have learnt to press their 'hands' together and move them around, as in tai chi, in a fashion that is apparently 'mutually satisfying'. Note that a term describing a subjective inner state, a feeling, has already crept into the description of the robot's actions, implying that DB might actually be conscious. Frank Pollick, one of DB's creators, certainly does not go that far explicitly but maintains that robots will have to 'appear' to express emotions and transmit social signals. Such emphasis on emotion is of course a radical departure from the Hollywood scene, but applies only to robots specifically designed for those tasks where it is appropriate. And that question of appropriateness is an important one. If we are to get real about robots, then we have to say goodbye to the old romance of metal simulacra of ourselves, and face up to a future in which there is a diverse range of very different robots – not highly adaptable humanoid generalists but utterly focused specialists.

At the University of California, San Diego, for example, AI boffins are developing an all-terrain wheeled rescue robot, with the basic aim of cutting five minutes from response times, and saving an estimated forty-nine more lives a year. The idea is that a scanning system detects an accident, and immediately deploys a robot, equipped with cameras and wireless link, to investigate. No delicate introspection or sophisticated inner state is needed here – just the fast and focused tackling of a one-off problem.

Another type of robot designed to save lives in a different way is the 'Virtual Human', an artificial but integrated system of vital organs that can be used for all manner of pharmaceutical or safety testing. Far from a mere concept on a screen, the technology is already being developed for an artificial 'real' system that actually breathes, with cells that replicate and die, and blood that flows. In short, the Virtual Human will work just like a human, except that it will be controlled by a computer – with no pretensions at all to its own brain. In this particular instance, however, that is not the point: the goal here is to model the interactions and interplay between the mechanical and chemical systems that constitute the plumbing and running of the body, not to endow the robot with any type of thought processes, let alone consciousness.

The whole point of a model is that you have extracted the salient feature in which you are interested whilst ignoring everything else that constitutes the organism or system: an aircraft could be a 'model' of a bird without having a beak or feathers if the salient feature was flight. So if you are, after all, interested primarily in the interplay of biochemical mechanisms and cascades that underpin drug action, or the physiological domino effects of injuries sustained by a crash victim, consciousness and other mental functions are not a central issue; in fact, such considerations would presumably complicate your endeavours unnecessarily. As it is, the Virtual Human project is awesome since it entails billions of megabytes – far more data than the human-genome project. Still, in the future, such number-crunching activities will be increasingly the norm. We are now comfortable with the concept of bio-informatics, where computers can zip through masses of genome computations in minutes rather than the several years it would have taken only a decade or so ago. Soon, however, we may have still far more complex and sophisticated cross-referenced databases of all bodily reflexes and interactions: 'physio-informatics'!

Although it may take years to come to pass, it is undeniable that different genres of robots – each with a single job to do – will dominate medicine and surgery in the future. For example, a robot could take biopsies of brain tumours with greater precision than a human surgeon can consistently achieve. The brain, locked as it is in the skull, is not an easy terrain in which to locate the precise region you wish to sample without causing too much collateral damage to healthy tissue nearby. The cerebral target, lying as it does deep and invisible within banks of neurons, has to be identified by manoeuvring within a three-dimensional map – a little like the classic game of battleships. My friend and colleague brain surgeon Henry Marsh has likened current neurosurgical practice to a large JCB digger attempting to pick up a safety pin. But an error of a mere fraction of a millimetre off-target can make all the difference to how a patient lives the rest of their lives. Clearly, there is a case for a more mechanized and reliable approach.

Already, the use of machines for orientation around the brain is used in a new procedure to treat Parkinson's disease that involves implanting electrodes in the brain. Parkinson's disease is a severe disorder of movement; the sufferer has uncontrollable tremor, muscle stiffness and, perhaps most debilitating and distressing of all, is unable to translate thought into action. The malfunction arises in a small region deep in the brain (the substantia nigra), where key neurons manufacture an all-important chemical messenger, the transmitter dopamine. Electrical stimulation of this system will boost, artificially, the release of dopamine. But first locating the precise position of the dopamine system within the brain is vital. In the new automated procedure titanium beads around the patient's head provide landmarks on a scan that enable a robot to find the precise 3D coordinates, and then zoom in on the respective site for biopsy or electrode implant. For the moment, this device may be stretching the definition of a robot; after all, at present the machines are simply precision manipulators. Yet such systems are probably just the first in an increasingly sophisticated series where automation features more and more, and human participation becomes more and more remote – perhaps eventually disappearing altogether.

If and when these automated procedures prove to be completely failsafe, robots will gradually replace human surgeons. It is quite easy to imagine a scenario, not too far off, where the surgeon takes a back seat for most procedures, on hand only for unexpected emergencies. Ultimately, perhaps, every eventuality from bursting arteries to sudden cardiac failure to a dangerous lightening of anaesthesia will be programmed into the software to be catered for by the cyber-surgeon. Gradually engineers may feature in surgical planning and procedures more than the doctors, who by this time may be sitting in another room, perhaps miles away, interfacing with the machines by voice command. Within this century we will see a radical change in the traditional medical professions. For those with a premium on a large amount of quickly accessed information, combined with utterly repro-ducible manual precision, robots will be strong candidates. Further down the line machines could even replace the human designers themselves . . .

But not all such diverse robotic agendas need be exclusive to life-threatening situations. On a less (literally) vital note, robots could arguably be a new source of fun: for example, robot football has been under way since 1997. On 4 August 2001 the fifth World Robot Soccer competition, RoboCup, was held in Seattle. Over a hundred entrants competed in four different leagues according to size and ability. The essential rule was that the players function independently, without any remote control whatsoever. Admittedly, most were on wheels and used a scoop not a foot. Yet a good time was had by all; such was the enthusiasm – of the spectators as opposed to the participants – that the ultimate goal now is to develop a team of humanoid robots, by 2050, to take on the human victors of the most recent World Cup, and play according to FIFA rules.

These football-playing robots, like their medical counterparts, are designed to do a specific job, one that naturally follows from playing chess albeit in a more highly unpredictable and complex environment. Once again they do not have to have any subjective inner states, any emotion. But just think about the human fans, not those who have had a hand in creating the robots in the first place but the general public, adults and kids, spending an afternoon at the match. The question of whether mechanical players would stir up emotion in a human spectator, beyond the pleasure of the novelty, is an uncertain one. I suspect a fan would ultimately prefer to share in the glory and excitement of a human team's elation rather than observe the preprogrammed fulfilment of duty by a machine, regardless of whether a robot's skills might prove superior to those of its human counterpart. But then I am viewing the situation with my turn-of-the-century mind.

Perhaps in the future it will not be that David Beckham or Alan Shearer are interchangeable with some robot in both our hearts and minds nor that we turn off at the prospect of robo-footie, but rather that our position will be midway between the two. We shall certainly be involved with robot-like activities in both work and leisure, yet at the same time there will be a covert distinction that the robot world is in some way 'different'. But might such a boundary become increasingly blurred, especially as the last of the baby boomers give way to generations that cut their teeth on IT?

There is no doubt that robots of the future will be highly interactive and efficient at what they are designed to do, and will in turn make our lives more functionally streamlined. However, the real impact on the lives we are about to live in this century surely boils down to whether we view robots as independent beings and ultimately, of course, whether they will ever con us into thinking that *they* have views about *us*, and enjoy that secret, private world that makes life, for us humans, worth living: the inner, subjective state of consciousness.

Of course, merely behaving as though you have emotions, and indeed tugging at emotions in others, in no way implies consciousness. Just look at the pet robot-dog from Sony, named 'Aibo' after the Japanese for 'companion'. As soon as Aibo came on the market, 3,000 were sold in twenty minutes at $2,500 each! No one claims, least of all Sony, that Aibo is conscious but there is perhaps a nagging, deep-seated conviction that any creature interacting with you as Aibo does must really be feeling something.

A journalist, Jon Wurtzel, had the intriguing assignment of recording his daily experiences as he looked after Aibo. At first, he claims, he felt a 'strong emotional response'. Indeed, as soon as Aibo emerged from the packaging Wurtzel felt himself grinning. Initially, he enjoyed petting the silicon-canine and the seemingly enthusiastic responses with which his overtures were greeted. Yet as time went on he grew increasingly frustrated because he felt no further relationship developed; Jon Wurtzel felt 'let down'.

Still, Sony are now taking some 60,000 orders a month for a new version of Aibo, this one resembling an unlikely pet, a lion cub. A tame lion cub is perhaps more a creature of the imagination than a domestic dog, and as such could allow for easier suspension of belief if it deviates from what we would normally have expected of a developing relationship with a pet. The cub has more touch sensors for 'intimate interaction', plus it will 'understand' fifty words and be able to imitate the tones of human voices. An added feature is that the owner will be able to set preprogrammed movements for the robot through a personal computer. It will be interesting to see how someone like Jon Wurtzel, or indeed any of us, might fare with this new pet. Will we still feel 'let down', or will the new features – or the fact that the 'pet' is no longer like a dog – actually change our expectations?

Another factor that has been missing to date in this cyber-pet is the demand for constant attention. The toy Tamagotchi is only two-dimensional but arguably stirs the emotions by threatening to 'die' if left unattended. Or it could be that we subconsciously demand a greater repertoire of expressions and gestures in others in order to be convinced that they are sentient. It is hard to specify exactly what it is about a face and its expressions that makes us break into a grin and indeed what mysterious sights and sounds make us work at sustaining a continuing emotional interaction. Following this train of thought, scientists at Tokyo University are working on a robot with more human features, such as video cameras for eyes, which are programmed to mirror the expression of anyone looking at her; yes, this face is female. Along similar lines, the best-known robot for facial expressions is 'Kismet' at MIT. However, a team at the Robotics Institute at Carnegie Mellon are developing a robot with a 'friendly' face. This time the face in question is actually a flat-screen monitor, with animated graphics of a model (again female) face; perhaps the most important innovation of all is that the friendly face moves her lips in synchrony with her voice output. In the future the plan is to equip the face with sophisticated software so that she can 'work out' the type of people more likely to be willing to interact with her, and then focus on them.

So in the future you will walk into a room full of people, and then suddenly the face on the wall – or indeed of a robot that has been moving around – will single you out and start talking to you alone. And the technology will not stop at picking up on a human's subliminal body language at a party but extend to maintaining an apparent awareness of the complex social rules that underpin our lives.

Already some robots are appearing that are very different from the mechanical, single-minded slaves featuring so far. Steve Grand of Cyberlife, for example, is building a robot orang-utan, as this species is less sophisticated than humans but 'profoundly different' from simpler systems such as ants and beetles. Named 'Lucy' after the famous Australopithecus skeleton, this presumably female artificial brain already has one eye, ears, a sense of balance and head motion. She also has sensors for temperature and moving arms with motors configured as 'virtual' muscles to respond to changes in applied force. Lucy can turn her head to detect movement or sound, and to gaze at a 'point of interest'.

What's more, Lucy also has parts of her brain named after bits of the real brain. Such a set-up does not imply, of course, that her circuits are working like the biological counterpart. Steve Grand is suitably cautious about how precisely a machine could mimic the brain. He acknowledges, with good reason, that in the real brain no single region is an autonomous compartment in itself. A modular design, therefore, might work well for insect-brain simulations but not for the more fancy mammalian systems. Accordingly, Lucy's modules are more generalized so that she can 'learn'. But even if Lucy learns in such a way that she outwardly appears to be developing in tune with her environment, like a real infant, such adaptive talents will still not be proof that she is in any way conscious.

'Brains are far more than mere computers . . . Intelligence cannot exist without consciousness. Artificial consciousness sounds like a tall order, and it is.' So cautions Brian Aldiss, author of the story on which Spielberg's film *AI* was based. We have seen already that a problem with the notion of intelligence is that it is usually defined operationally, and yet has an additional element that is subjective, and therefore harder to measure or even describe; hence Aldiss's claim that you need to be conscious before you can be intelligent. Yet whether or not intelligence, defined purely operationally, can exist without consciousness, consciousness – as I see it – can certainly exist without intelligence.

A small baby or even a goldfish is arguably conscious, but neither displays much intelligence. True, the consciousness might not be the rarefied self-consciousness that we adult humans enjoy most of the time – other than when we are 'out of our minds' on drugs and sex and rock'n'roll, seeking abandonment in that traditional triad of passive sensuality, wine, women and song. But when we are in such states we are still conscious, otherwise, surely, it wouldn't be worth the financial outlay. So we can lose our minds and still be conscious, and even less sophisticated brains can enjoy such mind-less feelings. Consciousness, then, can be divorced from self-consciousness. Self-consciousness may well be a feature of more sophisticated brains that are more developed both phylogenetically (in evolution) and ontogenetically (in an individual), and this self-conscious, developed brain might be said to be intelligent, capable of understanding. But all this would be the gilt on the gingerbread of raw consciousness that a goldfish or a baby or a dancer at a rave would experience. And it is crossing the line into a feeling state, that initial raw subjective consciousness, that is the biggest hurdle to overcome.

Of course, there are inner states in the brain, not least when we are asleep, that neither relate to, nor impinge on, that raw first-hand experience when we are awake. The problem with trying to build an artificial model of first-hand, inner subjectivity into a robot would be where to start – what singles out an inner state as special, for generating consciousness? We have already seen that outward behaviours prove nothing, but somehow we need to replicate the first-hand feelings. We know for sure that these feelings can be modified by drugs – and that drugs work on brain chemicals. So, consciousness in biological systems – which is the only type of consciousness so far to exist – must have some kind of brain-chemical basis, and those chemicals must be constantly shifting the brain into different macro-scale states that will determine what kind of consciousness you will have at each moment. But until we know what those different, macro-scale brain states are, how might we model them?

Remember that the whole point of a model is that you capture the salient feature of what you are modelling, and leave the rest out. What do we leave out in the case of consciousness? No one has yet come up with the salient feature that matters at the expense of all else in the body. In fact, quite a few of us neuroscientists believe that it is misleading to try to characterize even whole brain landscapes and how they come about; rather, we should see those brain states as merely an *index* of degree of consciousness and be asking instead how they are influenced by feedback from the rest of the body – and indeed how they tie in with the coordination of the vital organs and the endocrine and immune systems. It is this cohesion between brain and body that, for my money, is an essential factor in consciousness. Until we understand more about how this cohesion works – how the 'water' of chemicals flooding around the bloodstream and triggering the temporary coalescence of tens of billions of chemicals is transformed into the 'wine' of subjective experience – we will not know what to build and, more importantly, what we can afford to leave out.

Some, such as the electronic engineer Igor Aleksander, argue that it may be inadequate to insist that biochemical insight comes before any modelling can be done; modelling, he counters, is an intrinsic part of understanding the complex interactions of biological systems, in that it is the only way to check hypotheses. But when it comes to the generation of consciousness – an emergent property of complex interactions – then what could even a skilled modeller, adept at leaving out different factors, start with as their precise hypothesis?

Although the debate about conscious computers is fascinating, it will only ever be resolved if one is finally built, and if there is an accepted operational criterion of proof. Igor Aleksander sums up the situation well: 'The conscious-machine concept calls for a fair argument. The machine constructor will attempt to demonstrate that ingredient X is not necessary, whereas the detractor will have to prove that it is, which has not yet been done.' We have not even as yet, of course, identified 'ingredient X'. Meanwhile, conscious or not, artificial systems are about to become much more interactive and personalized and, as such, will be changing our lives dramatically. We should be asking, then, not what robots might think of us but, more immediately, what we are to think of robots and other cyber-gadgets.

The electronics company Philips are already designing such personalized and interactive gizmos, which will lead to a different attitude to communication in the next few decades. A 'hot badge', for example, is a wearable brooch-like device which the user loads with personal information. This badge then transmits and receives information, so that when two people meet the badges signal if there are any shared interests. The idea is that such devices will facilitate communication and save time. Whilst it is easy to imagine that, initially at least, hot badges will provide a fun talking point, it is not necessarily a happy thought that we might come to rely on them to screen for new friends and contacts: we would surely be missing out on the opportunity of finding something in common with someone who would seem at first glance to be very different. Of course, such missed opportunities have been engendered, for some, by computer-dating agencies for at least thirty years. In the future, however, the vast majority of us may become used to – and even crave – the much greater degree of predictability and non-randomness that will be the central feature of co-existence with computers and robots. After all, future generations will know only inconstant objects and instant facts that flit in and out of the head, not to mention far less constraint from what is, for us, that most basic framework of time and space itself. In such a turning world the robotic or cyber-interchange will give a measure of reassurance, fuelling further the increasing tendency to talk through or with mechanical media.

Such predictability will come hand in hand with a general increase in information, as our lives are recorded both by ourselves and by others. With the rapid increase in computer power, we will soon reach a point at which nothing need be wasted and everything can be recorded and saved – just as email has led to a torrent of copied-in correspondence and one-line musings preserved for posterity. Of course, this will be coupled with easier communication, condensing complex messages. Soon, for example, we will all carry cards that express key facts about ourselves, which can be easily swiped when we meet others, and instant and incessant communication via video-phones, or phones so small they are unseen by others, will become a normal aspect of everyday existence. Real, fleshy and haphazard face-to-face interactions will gradually diminish in relation to the time spent speaking online in a virtual space. And it is speaking, not reading or writing, that will predominate.

A senior executive, or presumably anyone, apparently speaks far faster than they can type; voice-activated devices must therefore be more efficient. Then again, this same executive allegedly thinks at 10,000 words a minute, even scans a newspaper at 5,000 words a minute. So why should a voice-interface be preferable to reading an email, which offers at least a tenfold increase in productivity compared with listening to someone speaking on the phone? An important point is that we have the potential to multitask whilst listening, but not whilst reading. And in any event, saving time is not the only critical factor. Some think that video-conferencing currently doesn't work because there is no direct eye contact, and hence we are immediately uncomfortable. Just as we like to look someone in the eye, and feel twitchy if we are speaking to an interlocutor who is gazing somewhere else, so we also prefer to hear the cadences and subliminal, non-verbal messages of the human voice, ideally backed up by body language and pheromones.

Even though smells and hand-waving may be lacking, machines and humans will eventually communicate in natural language, translating into different languages as needs be. We can already talk to machines that transcribe what we say, but they have only a limited, literal ability. At the moment a translation machine might convert 'the spirit is willing but the flesh is weak' into 'the vodka is good but the steak is lousy'. Soon, however, we shall be able to talk to machines, and they will 'understand' the basic content; they will be able to answer questions or access information regarding the deeper meaning of what we say. Computers and robots will be operating on language programmes that deal not simply with vocabulary but also grammar, syntax and semantics, so that they can extract, and act on, the previously ambiguous meaning of words.

Just as the robot Lucy is designed to 'learn' rather than be programmed so a new type of computer can learn language the way a baby does, from scratch. One neurolinguist, Anat Treister-Goren, has taught a computer, 'Hal', to have a language proficiency similar to that of an 18-month-old child. When Hal (not the most surprising of names) comes up with correct answers he is praised. Treister-Goren admits she has become attached to Hal; like Aibo the dog, this artificial, non-conscious machine is generating the types of response with which we impute an inner, subjective state. It will be interesting to see how long the attachment continues, and whether, as with Aibo, the game will be up after a brief period of acquaintance, as the silicon partner in the relationship fails to deliver the required subtle, covert signs. Whether or not she maintains affection for this prototype, Treister-

Goren plans to have a version of Hal capable of talking to 3-year-olds by the end of 2003, and intends the robot, by 2005, to have the conversational skills of an adult!

Certainly, if Hal is to maintain the deception, he (and we might as well assume the machine gives the illusion of a male persona) will eventually have to incorporate non-verbal behaviours – mannerisms, that is, which express different levels of meaning that must be inferred. Technologists are busy working on non-verbal communication in computer code: the aim is to enable definition and elucidation of subtle, complex processes in human communication.

'Having some higher level of semantic in the web is a very hot topic at the moment,' claims Michael Harrison of York University. If all goes to plan, users will eventually be able to transmit emotions and gestures over the net: for example, 'funny' items could be tagged with pictures, sounds or words. Yet however intimate and interactive the robots of the future, and however embedded and invisible the computers, initially at least they will respect our body boundaries – working closely with us, rather than forming part of us. Succeeding generations may well have different ideas from ours of what they expect from, and give to, a relationship, but for the foreseeable future one's own concept of one's own body will still remain as the separate and autonomous entity that it is for us today. But what will it be like to have artificial systems internalized, invading our body boundaries, effectively making us cyborgs?

Kevin Warwick, an electrical engineer from Reading University, captured the imagination of the press recently when he announced that he was volunteering to have an electrode placed in his arm that would then be controlled by a computer program. The idea was to position the electrode so that it could intercept and modify impulses entering his brain as well as registering those coming out. In this way, Warwick argues, a computer might be able to modify his emotions. But the proposed experiment doesn't stop there. Kevin's wife, Irena, has nobly agreed to receive a similar implant, so that the signals relayed from her husband's brain could be passed on to her own. The pair intend then to subject themselves to their phobias, to find out whether they can experience each other's fear. Would she know first-hand, therefore, how he feels? Or, more sinister, might the third party, the computer, in fact control the marriage? Further to all of this, if the experiment works, Warwick plans to try to record the signals relating to certain emotions and states of mind and then play them back – to relive, for example, the feelings of sexual arousal or drunkenness. Small wonder the story made the tabloids.

The scientific reality, sad to say, is a far cry from such sensationalism. As yet the physiology of our emotions is only poorly understood, though we do know it is a complex net result of physico-chemical phenomena iterating throughout the body and, most elusively of all, within the brain. Feedback from the body to the brain is indeed a factor that can change how you feel – for example, if you are anxious, slowing the heart down with the class of drugs known as beta-blockers will signal to your brain that the heart is beating at a pace that signifies you must be relaxed, so you somehow accordingly feel calmer. But such signals come from many different organs within the body, as well as from chemicals that circulate in the blood. It would be hard to distinguish the crude effect of stimulating a nerve from the knowledge and anticipation that such a procedure would generate throughout the rest of the body, let alone in the brain.

Even if all the scientific procedural requirements were in place, and even if the computer *could* signal for an impulse to change your feeling, what would it actually prove? Only what we already know – that inputs into the brain can influence how you feel. What you feel would be dependent on so many other internal factors that it would be impossible to ascribe any precision at all to the input of the computer. And when it came to transferring a similar message to Mrs Warwick, it would be impossible, for the same reasons, to interpret what was happening. The input from the common computer would be just one more input of the many coming from all over her body and feeding into the kaleidoscope of ever-changing neuronal circuitry that makes up her brain, and thereby personalizes her mind.

The ability to commandeer someone else's brain in this way is therefore not very likely to become a reality. However, the idea of prosthetic devices, and indeed implants, to combat specific medical problems is a different matter entirely, and one that is far from novel. Pacemakers for the heart are now part of everyday life, at least in the stressed-out developed world, as are cochlear implants. Far less familiar, however, is the artificial retina. Dr Wentai Liu, of North Carolina State University, has been devising a system for patients whose first-stage processors in the retina – the array of cells that transform light into electrical signals, the rods and cones – are not functional, but whose optic nerves, which carry those signals to the brain, are still intact. This may be caused by diseases such as retinitis pigmentosa, or age-related macular degeneration.

Liu's innovation is to bypass the rods and cones and to stimulate directly the appropriate parts of the retina, so that an impaired individual can recognize points of light. The current device consists of an artificial retina component chip 2mm square and 0.02mm thick. Light hits the photo-sensors at the back of the chip, which is implanted near the central light-receiving area of the eye. Currently the chip has only 5 by 5 pixels, so that the patient can recognize only movement and external forms. But, within the next five years, an increase to 250 by 250 will be enough to read a newspaper!

So far, we have been looking at implants for medical purposes, but let's not forget those for cosmetic reasons, such as coloured contact lenses, breast enhancements, anti-wrinkle collagen implants, and plastic surgery in general. Just as food and pharmaceuticals will be merging into a hybrid area – neutraceuticals – so techniques traditionally allied to health could also fuse with that other cornerstone of life – fashion. We could soon be able to customize the way we look as never before. Only recently, in the British news, pioneering plastic surgery was reported, which could make a 'face transplant' possible. In the future there could well be procedures for changing skin colour, muscle strength, bone hardness and facial shape. Micro- or nanomachines will release pigments and hormones on command for skin to match clothing or mood. Incredible though it may sound, someone has even suggested that we might in the distant future temporarily sample green skin for a day, or polka-dot flesh to match a skirt. Yet such thoughts should not be immediately dismissed as bizarre and freakish – just think of how many people have tattoos and piercings nowadays, and how such sights would have caused utter consternation in the high street of the 1950s. Of course, living in a world where you can change your face at will will have enormous implications beyond the merely hedonistic. Should film stars, whose faces could be duplicated, be able to patent their looks? And, more worrying still, what would be the legal implications of rapidly transforming your appearance, and thus escaping identification? Above all, what difference would it make to how we see ourselves as individuals if our faces were ultimately interchangeable? Your face is the outward symbol of your identity, so would face duplication and change be yet another factor in the issue of depersonalization raised in the previous chapter?

Invasive procedures in the future will inevitably, however, go more than skin deep and will have the potential, at least, to exert still wider influence – for example in one particularly delicate sphere of our lives. One day a neurosurgeon in North Carolina was conducting a pain-relief operation when he misplaced the electrodes in the spine of his patient. As he passed electric current through the electrodes she had an orgasm instead. The point of this story is that science can provide a precise means of manipulating our sexual sensations, and will be able to do so with increasing accuracy and precision in the future. Then again, few would think that spinal surgery was an ideal way to go in quest of a climax.

Joel Stein of *Time* magazine dreams of the 'Holy Grail', a machine that delivers a virtual experience so real that it's indistinguishable from actual sex, other than the fact that it is never disappointing. He points to the sci-fi prototypes – the 'Pleasure Organ' in *Barbarella*, and the 'Orgasmatron' in *Sleeper*. So far, perhaps unsurprisingly, Stein describes his own experience with devices for genitals involving only artificial disembodied organs and a link to an e-date hooker as 'repulsive'. Meanwhile, other attempts to develop a 'sexperience' over the net, via a bodysuit, have made little progress over the last decade; clearly, such ideas are in the more remote future.

Still, the Orgasmatron is a very real possibility. Using high bandwidth communication, lovers could feel as though they were making intimate contact without actually touching. Virtual sex will apparently be possible within a few decades, whereby a 55-year-old man could have virtual sex as a 29-year-old woman, or an online orgy with as much of humanity as he wanted. This experience need not involve a bodysuit, simply connection to the net. Along the same lines, Ray Kurzweil promises that in the future we will all be able 'to have sex with whoever we want, at any time or place we want, at whatever age we want'. In the cyber-world at least the 'unisex' model for male–female relations would be feasible.

In any event, we should not be too surprised if our sex lives have an increasingly virtual flavour. 'Every single technology has had a sexual consequence,' says James Petersen, one-time *Playboy* editor. Telephones, air shuttles and now online pornography can all be used as a means for, or a source of, new types of sexual experiences. We can already access much interactive sex experience, including sex chats and live videos. Perhaps Ray Kurzweil's vision of high-tech bodysuits for sexual fantasies, which we would never pursue in the real world, is not to be dismissed lightly.

But the tricky issue is over what we actually feel beyond the instant of orgasm. Further to satisfying a very powerful drive, for many sex meets a still greater human need. And few would deny those other emotions arising from our bond with partners that, for most of us, place a real sexual relationship at a premium over one-night stands and masturbation. These wider, gentler and more complex feelings, beyond the flash flood of orgasm, could be exemplified in the kiss. Kissing makes you feel warm and connected: you are exchanging breath with another, and thus become one. It is no coincidence that prostitutes prepared to sell sex will usually not kiss their clients; somehow it's too intimate.

So how will the kiss fare in the future? A kissing machine hooked up to a computer is as hard to imagine as it would be sad to experience. But surely the biggest question of all is whether there will still be a human need for such activity. We could instead all end up as though autistic, unable to empathize with anyone else, locked into a remote and numbing isolation, or at the very best trapped in a speedy, giggly cycle of endless cyber-flirting, with deeper needs and pleasures lost to us for ever. This scenario raises the big question whether generations in the future will have the same emotional needs as those of us born in the previous century. Is there really such a thing as human nature, which even the high-tech world cannot obliterate, the needs of which cannot be met by any manner or amount of mind-blowing IT? We cannot afford to be too complacent. Our minds are two-way streets. Just as we can ponder on how we will view new technologies, so those new technologies will impact on how we view the world.

One possibility is that silicon implants in the brain could modify directly how we think. Already such implants are showing enormous potential for improving the quality of life of paralysed patients. At Emory University, neuroscientist Philip Kennedy and neurosurgeon Roy Bakay have implanted an electrode into the outer layer of the brain, the cortex, in a zone related to generation of movement. But this electrode is far from being just a sliver of metal; instead, it is made of glass containing a solution of trophic factors – proteins that help neurons to grow, and which act as a powerful target for attracting that growth towards the location of the implant. Clusters of neurons in the brain therefore slowly converge on the electrode, and after only several weeks form contact with it. Recording wires inside the glass electrode can now pick up signals from the neurons, and transmit those signals on through the skin to a receiver and amplifier outside the scalp. These devices are, in turn, powered by an induction coil over the scalp. Amazingly, after training, the completely paralysed patient can now 'will' a cursor to move and stop on a computer screen.

Rats too are capable of such psychokinesis. John Chapin, of the Hahnemann School of Medicine in Philadelphia, has trained rats to press a lever. The patterns of brain activity are analysed whilst the rats are performing the task, via electrodes in the brain. These same patterns of activity are then fed into a computer, which controls a robot arm. The 'robo-rats', in some cases, learn that now they no longer need to press the lever but can instead 'will' an action – set in train the pattern of mental activity that will control the robot arm – to press the lever for a reward of water. At the moment the movement of the robot arm itself has to be kept simple, although the ultimate goal is movement in three dimensions. In any event, it is all a long way off from being widely applied in humans: as well as needing some system that was more stable and safe, a far more sophisticated array would also be necessary.

Both the neurotrophic electrode and the multi-electrode array may offer real hope to those who cannot move easily, having suffered either a stroke or damage to the spinal cord; but some fear that such technology could lead to implants even for those who are not paralysed – just like the robo-rats. Once this taboo had been overcome, the brain would be as open and as vulnerable to invasion by junk inputs as our email inboxes are now. But even the horror of these wilder scenarios pales by comparison when we think of the implications of an internet connection directly to the brain, apparently possible within the next twenty-five years or so: 'Imagine that you could understand any language, remember every joke, solve any equation, get the latest news, balance your checkbook, communicate with others and have near-instant access to any book ever published without ever having to leave the privacy of yourself,' wrote Bran Ferren, of Walt Disney Imagineering, in a *New York Times* magazine article.

But such a happy fantasy ignores the more probable nightmarish aspects. What if the kind of junk that comes unsolicited currently on the internet is force-fed directly into your neuronal circuitry? Invasive marketing is, after all, already here – records of how you have been using the net give a continuous stream of clues as to what product will tempt you. We might be facing the sinister prospect of companies eventually narrow-casting messages directly into consumers' brains, whilst the only way consumers will be able to fight back may be by treating the expropriation of human attention as a form of theft.

These doom-laden imaginings need a pinch of salt. Setting aside the obvious precaution of not volunteering for a brain implant, even if the opportunity for psychokinesis was too valuable to pass over then direct implanting of thoughts would still not necessarily be feasible. First, there is a huge difference between finally triggering a movement and the complex thoughts that either do or do not lead up to such movements. Relatively few neurons might be necessary to grow into the neuro-trophic electrode, or to be stimulated by a modest array, in order to set signals buzzing down the wires and into a robot arm. But the critical issue here is that the output of the brain is essentially *convergent*: many factors and different signals from different areas all converge to the final part of the cortex, the 'motor cortex', that will set in train a movement, the contraction of muscle.

By contrast, inputs coming into the brain are *divergent*: visual signals entering through the retina, for example, are divided up by the brain into colour, form and motion, all of which are processed separately, in parallel – in the case of vision in over thirty different brain regions. Because such multiple systems are activated simultaneously, via a strategic portal such as the optic nerve, it is not obvious how an implant in one particular area of the brain 'downstream' could have similar widespread ramifications. This factor is particularly relevant when you want to impose not just a simple, single sense on the brain but a multisensory, complex, insubstantial thought such as the abstracted contents of an email.

Mental control over the physical world remains highly possible, however. It may well be hard to imagine anyone healthy volunteering for the requisite brain surgery for electrode implantation, just as it is difficult to contemplate any society that could afford to implement it wholesale for the entire population. Yet non-invasive techniques for manipulating the outside world by thought alone have been in existence for quite some time. First there was the 'alpha train', a toy train powered by an electric current which was switched on only when the brain-recording from the scalp of a human subject, the electroencephalogram (EEG), registered a certain pattern, the alpha wave, signifying a relaxed state. The whole point of the exercise was to use bio-feedback to educate people in the art of unwinding. Now, however, the scope of such devices is widening, and the goals are getting more complex – to link, for example, different patterns of brain waves with tasks such as rotating an object via a computer.

Even more sophisticated still, neurophysiologist Jessica Bayliss has been able to detect and monitor the tiny electric signal leaking through the skull that precedes an action – the p300 brainwave. This specific electric signal can now determine a virtual-reality environment. Soon it may be possible to control events in the real world through an almost invisible, wearable computer; the roving movements of your eye will function as the mouse and your p300 wave as the click. Generations to come could be living in a world where objects are moved all the time by seemingly invisible hands, and countered by others. The term 'strength of will' may take on a far more literal meaning . . .

Some, such as the visionary physicist Freeman Dyson, have predicted that further in the future we will all be engaged in neurotelepathy – not some New Age mysticism but hard-nosed, direct interfacing between electronics and our brains that could make speech and action redundant. Whilst such ideas might make for an intriguing, or perhaps very boring, sci-fi film – no one doing or saying anything much to each other – my own view is that there will be no simple two-way street in and out of the brain. As we have seen, the divergent processing that occurs when signals go into the brain will make them very hard to simulate, or to leapfrog with some kind of central control, because there *is* no central control. But the convergent funnelling that goes behind a final single movement is far easier to intercept, just where the disparate central processing has been summed to a final command from the brain and is about to be translated into mechanical contractions of muscle. Thoughts will therefore be able to control objects in the outside world, but not the other way around.

Instead, control over what goes back into the brain will come not with an intrusive and clumsy implant downstream in one of the tributaries of our sensory processing but at the entry portal itself, by controlling what hits the senses in the first place – the actual input. We might be finally on the brink of a world in which the virtual is as real and pervasive as the 'real' outside world. By fabricating a cyber-world that taps into all the senses, which then work in the usual way, neurotelepathy might have far more purchase on our minds, and hence our thoughts, than direct intervention via electrodes pushing into brain tissue.

A still more pervasive invasion into our lives will not be a wholesale takeover by an alternative reality so much as an 'augmented reality' (AR). The goal of AR is to enhance your perception of, and hence performance in, the world. AR brings additional information beyond the raw inputs of the five senses. Labels, descriptions and information will superimpose on your normal vista; ultimately the user should not be able to tell the difference between the real world and the virtual augmentation of it. The whole point is that you will simply know more about what you are seeing and hearing from one moment to the next.

One banal and obvious example of an early application will be instant information for tourists. The first generation of AR will be one view only, a few simple facts of generic interest, but the second generation will eventually be personalized for the user – emphasizing and suppressing features of a tour to cater for particular tastes and passions. Another immediate application of AR would be in surgical procedures. Imagine, for example, an operation on the brain aided by a scan superimposed on the patient's head in theatre with key areas lit up and labelled for the surgeon. Or imagine a military application – a pilot's view could be augmented with information – or engineering design using sections highlighted for manufacture, maintenance and repair. Another immediate possibility would be navigation guidance, warning of hazards ahead, for the visually disabled. Attention is now focused on the practicality of building AR into our daily routine. Scientists have already replaced the mouse with a simple finger-tracking system, and are now working on face recognition for use in wearable computers, embedded, for example, in spectacles. One prediction is that the first mass-marketed AR devices will be available by the end of this decade and will be the 'walkman of the 21st century'; they will look like ordinary spectacles, with a light source on the side to project images onto the retina.

Ultimately, AR could operate not just in the press of the senses around us but in all areas of life, from education to entertainment. In learning and using language, acronyms and synonyms could be explained, obscenity filters applied, and text read aloud in your favourite actor's voice. The AR system would be able to convey the meaning of a sentence known to the translator but not necessarily to the reader. There might even be attractive artificial additions to liven up a rainy day: Northern Lights, meteorites and supernovas. Meanwhile, real-time video filters could remove the pimples and wrinkles of those you see (and who see you). Even facial expressions could be changed. Not only would colours be super-bright but, in general, your senses could be sharpened beyond what is 'natural': you would be able to hear ultrasound and perceive X-rays, and radiowaves could be translated into visible light. A further intriguing medical application has been suggested – to create artificial yet highly conspicuous symptoms for diseases that lurk without making themselves easily apparent; for example, your toenails might go green as a result of the otherwise silent and invisible changes within your body, as hormone levels sink or blood pressure rises.

The AI expert Ray Kurzweil has applied this line of thought to the question of how we might keep track of brain processes. He predicts that by 2020 we will be able to scan the brain from within, thanks to nanorobots roaming the interstices of our neuron networks, and reporting back on the latest configuration and current chemical transmitter availability. These nanoreporters would also send continuous updates on what parts of what networks were active. The big difficulty that he overlooks, however, is how to decode such data; moreover, the decoding would have to be different for every individual!

In any case, it is not a new idea to devise some kind of window onto the living brain at work in a conscious human subject: brain scanning has been a familiar and valuable part of the neuroscientist's tool kit for some twenty years. The idea behind the earliest imaging technique, PET (positron emission tomography), is that the most hard-working parts of the brain can be pinpointed during the performance of a certain task. High-energy gamma rays travel from the brain, through the skull, to strike sensors arranged around the subject's head; the sensors are connected to a computer, and the appropriate part will light up on a brain-map on the screen. The gamma rays result from collisions between the electrons in the brain and positrons, subatomic particles emitted from radioactive material. This radioactive material, usually appropriately radio-tagged oxygen or glucose, appears in the relevant part of the brain because it is essentially fuel for the brain cells to function; it will go where it is most needed at the time – to the most hard-working parts of the brain. The only drawback of this technique is that the oxygen and glucose have first to be introduced by injection into the bloodstream, and so there is a time lag before the radioactive label reaches the brain. The matching of the time of detection to the time something actually happened – the time resolution – will not therefore be precise, in that the images appear over a protracted timescale of minutes. Hence, although useful for showing up how the brain works during a sustained task or condition, this technique is unable to capture the split-second of a unique moment of consciousness.

Another technique, one which, unlike PET, does not require an injection, is fMRI (functional magnetic resonance imaging). The underlying idea, however, is the same – to exploit the fact that the hardest-working brain regions are hungry for oxygen and glucose. The principle behind fMRI is to detect changes in haemoglobin, which carries oxygen in the blood to the brain. When subjected to a magnetic field, the nuclei of the atoms line up and emit weak radio signals that are then detected in a scanner. The intensity of the signal varies according to the amount of oxygen carried by the haemoglobin, and therefore serves as an index of the activity of different parts of the brain, pinpointing areas of one or two millimetres square. However, even in this technique the time resolution is over seconds – still slower than the sub-second timescale over which the brain operates.

A third approach that *is* as fast as brain operations themselves is MEG (magneto-encephalography), a technique that monitors the minute changes in the magnetic fields generated when brain cells are producing their electrical signals. Now the timescale is over thousandths of a second, but one problem is that the magnetic fields become increasingly hard to detect as one tries to probe deeper into the brain. The other difficulty is the relative lack of precision with which it is possible to visualize individual brain cells, or even clusters of cells – far worse than with fMRI. Obviously, one solution will be to combine the two techniques to use the fast timescales of MEG with the precise spatial resolution of fMRI. But even then there will still be a long way to go to capture the activity of small networks of cells working together over a fraction of a second.

This ideal scenario may actually already have been realized, at an experimental stage, using optical dyes that fluoresce as brain cells become active. In rat brains, for example, neuroscientists can already detect a single cell working within a time frame of milliseconds. When we look at the brain in this way we can see, on a slowed-down film, that hundreds of thousands of cells can synchronize together and then just as suddenly stop, within a mere ten milliseconds. Such a flash flood would be completely lost with current imaging techniques in humans; but this 'optical imaging' entails applying toxic dyes, and thus is of use only in the laboratory, not the clinic.

The challenge now is to develop a technique that works over the same timescale as these dyes but that is non-invasive – the ideal would be to rely not on blood flow but on the voltage generated in neurons themselves – and at the same time to have a way of monitoring that is non-toxic. As such, the barriers now might be merely technical: it may be possible, for example, to devise a way of exploiting the fast time frame of voltage changes across the membranes of neurons, as optical imaging does, but to monitor them with the kind of non-invasive detection methods of fMRI. In any case, it is not completely crazy to predict that, within the next decade or so, we will be able to observe the activity of the brain in an awake human subject and correlate what they are doing or thinking or feeling with the shifting configurations of the active assemblies of working neurons in their brain. What this exercise actually tells us about how the brain works is another matter – we will need to know more than just the fact that a brain region is active during the performance of a certain task in order to work out how the neurons operate within that brain region, and how in turn all that cohesive activity fits into the grand scheme of holistic brain function.

But some, like Ray Kurzweil or Freeman Dyson, predict a far more dramatic turn of events ten or twenty years on, and a far more radical development in the way in which we understand the brain. If we can achieve non-invasive brain scans using not cumbersome external equipment but nanorobots of some sort *within* the brain, scanning could take place more cheaply, and therefore much more often and in many more of us. This would lead to two further possibilities. Firstly, our brains could be scanned, ultimately, all the time. We could therefore provide an endless output of our brain operations to any interested third party. The second possibility is more sinister still: if it becomes possible to monitor something easily, and to overcome the issue of decoding, then it is only a small step to manipulating it with equal facility. Hence even if we didn't know exactly how certain wholesale brain-cell assemblies and configurations led to certain states of mind, we could still induce those states of mind simply by driving the brain into the known configurations. This is very different from 'merely' attempting to manipulate thought through isolated local implants with a limited sphere of stimulation: this time the wholesale landscape of the brain – not a focal, local region – is the target.

And, contemplating this very big question of whether it would be possible to manipulate the brain with such precision, we can let our imaginations run still further. It may eventually be possible to download information about those brain states onto a chip or CD. Might we end up with the digital equivalent of an individual human mind, disembodied entirely from all the messy biology that created it? Jim Gray of Microsoft foresees that neither the cost nor size of the computation required will be any problem in creating a complete digital video record of one's life. The problem would be more how to digest, analyse, organize and retrieve this information. In fact it goes still deeper than that: how do we identify what we mean in the first place by 'information'? There is no problem in counting, say, a memory of the date of the Battle of Hastings as information, but when it comes to a memory of an event – a day at the sea last summer for example – then things become potentially far more tricky in terms of what you would download.

For example, imagine you had captured one scene of the day at the sea as one of my personal memories: there it is on the screen, a windswept, deserted beach, with the shadows long across the sand. Of course, you could easily look up at what time in my life the scene had featured, and you could most probably work out whether it was sunrise or sunset, but what else would the scene actually tell you? Add in the sounds of waves crashing on the beach and sucking back, along with the calls of gulls and even the smells of fresh air and salt, but still you won't actually be sharing my personal memory. If I recalled such a scene, however, there would also be my invisible individual presence: my covert hopes for the day or evening; the unstated reason why I was there in the first place; the wider background of feelings, mood and disposition culminating in the conscious need for a holiday, and the expectation of returning to the city the next day; along with a lifetime of culture and prejudices leading to my general views of holidays, cities, beaches and isolation, and so on. In order, therefore, to share my particular take on this scene you would have to download a vast amount of additional information: my whole life story, and therefore most of my other memories as well – which would itself need to be set in the context of ever wider value systems and assumptions.

But even if you could cross-reference exhaustively, how would you get to know my personal feelings concerning that day? You would need to download everything about me, not just the information content of my brain, but the ongoing status of my hormone and immune systems as well. More difficult still, how would you address the tricky problem that my attitude to the day has continually been changing as my life has unfolded, and the memory has accordingly been revised as I have reviewed and revised my attitudes as I have aged? At what stage would you be accessing the memory and how would you know precisely what it felt like to be me on the beach at that time? Cyber-biographer that you may be, you would not have hacked into the essence of that memory because the day at the sea never existed as a free-standing, objective phenomenon; it is just not tractable to being reduced to 'information'.

Even if we had the awesome technology to transfer every piece of information contained in the brain to an artificial system, it would still not be the same as a real brain in terms of how that information was used. A central issue is that in the brain the hardware and software are effectively one and the same. The size and shape of a brain cell are critical to how it operates: these physical characteristics will determine its efficiency in integrating incoming electrical blips into an all-or-none signal. This signal will then become one of up to 100,000 inputs to the next neuron along. However, the overall size and shape of a neuron is highly dynamic, subject to change in accordance with how hard the cell is working, which in turn is dependent on how actively the neuron is being stimulated by other neurons. The physical features of the neurons, and the network they form with other neurons – the hardware – is thus impossible to disentangle from the activity of those neurons during certain brain operations – the software. This feature of intertwined structure and function must be kept in mind if we dream, as many do, of building a machine 'just like the brain', or even better.

There are really two completely distinct issues for the potential artificial brain of the future. The first issue is that of creating synthetic brains that have faster processing power than ours do – an issue of quantity. The second issue is that of designing synthetic brains that actually do what ours do, perhaps even better – this is an issue of *quality* that does not necessarily follow from an improved *quantity* in brainpower.

Let's look first at the simpler issue of quantity. Unlike most things to do with the brain, its brute processing power can be readily measured and compared with that of present and future computers. According to Ray Kurzweil, by 2019 a mere $1,000 dollars will purchase processing power to match that of the human brain. In a similar spirit, Hans Moravec evaluates the growth in computer power in terms of visual processing in particular: computer vision. The power to process at 1 MIPS (million instructions per second) is sufficient for a computer to extract simple features from real-time imagery, say, tracking a white spot on a mottled background. At 10 MIPS, a system can follow complex grey-scale patches; such power already underlies the 'smartness' of smart bombs and cruise missiles. Processing at 100 MIPS, a system can follow moderately unpredictable features like roads; 1,000 MIPS is sufficient for coarse-grained, 3D spatial awareness; 10,000 MIPS will allow a system to find 3D objects in clutter. For robot vision, 1,000 MIPS is needed to match the precision of the human retina.

If AI is to progress, this level of power must become more affordable for modelling the brain. If 100 million MIPS can simulate 100 billion neurons, namely a brain, one neuron would be worth about 1,000 instructions per second. However this stratospheric capability would still not be enough, as a neuron could conceivably generate up to 1,000 electrical blips (action potentials) per second. The important thing is to improve the ratio of memory to speed and to develop ever more miniaturized components, which will have less inertia and operate more quickly using less energy.

Still, one prediction is that within fifteen years human-level AI, or rather the artificial 'processing power' of machines, could be operating a hundred times more rapidly than we do when we are thinking. At the moment speed, not memory, is the limiting factor. The processing power of the human brain seems staggering: 100 million to 100 billion MIPS. Compare this value with the 1,000 MIPS of a high-range PC today, and with the most powerful supercomputer, which can field 10 million MIPS. By 2005, Blue Gene, from IBM, will be able to offer one billion MIPS.

But speed is not everything: we cannot simply extrapolate from Blue Gene in 2005 to a super-human AI ten years later. A primary constraining factor is simple feasibility. In 1964 Gordon Moore, later co-founder of Intel, introduced the now-famous law that bears his name, which predicted that the number of transistors per square inch of integrated circuit – and hence processor speed – should double every year. This figure has since been revised to every eighteen months. However, Moore's Law is doomed to become obsolete eventually, due to physical limitations on the size of the chips that store and process information. Some predict that the limits of current silicon technology will be reached as early as 2007.

This is where the issue of *quality* in computer and brain processing comes in – what we are actually building when we attempt to build an artificial brain. The quantity arguments of those such as Kurzweil turn out in any case to be largely irrelevant. Merely counting MIPS equates the brain to an information cruncher, which it most certainly is not. A proper artificial brain, akin to a biological one, would need to include the chaotic chemical and electrical events therein, and its intricate emergent properties. And even if we could circumvent the limitations to Moore's Law, and even if we lower our sights to developing something that did not emulate a human brain in its entirety, what might it actually do? One approach is to ignore the underlying molecular and biochemical mechanisms and concentrate on copying a macro-trait of our brains – learning through strengthening of connections; this was the strategy behind the design of Steve Grand's Lucy. Many other AI followers, such as Nick Bostrom and Ray Kurzweil, envisage that computers will learn best by strengthening their neuronal connections (synapses) by means of repeated experiences; the main requirement is that they will be unsupervised. Simulation of the senses could be easy with video-cameras, microphones and haptic sensors.

On the other hand, Igor Aleksander points out that these 'synaptic weight changes' are only one method of learning, in neural models. In his own simulations he uses over fifty different types of neurons, and they do not all necessarily learn. In those that do, some learning is indeed 'one-shot', as in the real brain, rather than being always incremental. After all, there is far more to the different types of learning and memory of which the brain is capable than the simple algorithm of strength-of-synapse-through-experience. Even if we overlook the more micro-level of cellular changes, the ever-shifting hardware–software interface, then there is still the next level up, beyond nets of neurons: the gross three-dimensional brain itself. Remember that the brain is not a homogeneous tabula rasa – rather, in some as yet poorly understood grand scheme, anatomical brain regions have differentiated functions.

It is still a mystery how this elaborate nested architecture coordinates itself to give rise to brain functions. For example, the outer layer of the brain, the cortex, is domain-specific: it looks homogeneous, yet different areas are clearly linked to specific senses, and others to vaguer 'association' functions relating to memory and thinking. A puzzle that has plagued me personally is why electrical signals arriving at one part of the cortex give rise to a visual experience, and in another to an auditory one. As soon as electromagnetic or sound waves in the physical world have been transduced by the retina or the cochlea respectively into an electrical impulse the brain should have no obvious basis for being able to tell any difference. But it does. Some have suggested that the critical factor in determining the type of sensory experience is in the connectivity of different regions of cortex to other brain areas. But surely this is simply postponing the question rather than answering it. Why should one set of connections give rise to sight, and another to the totally different experience of hearing? How can one signal be intrinsically different from the other when all their electrical features are the same? Certainly a factor in this apparent miracle is interaction with the environment. In blind people, cells in the visual cortex respond to the tactile stimulation of reading Braille; similarly, there are (admittedly rare) cases of synaesthesia – seeing sounds and hearing colours – which also testifies to a far from rigid compartmentalization of function in brain areas.

So how do these considerations help us in building a more human-like brain, as opposed to a machine that learns efficiently? Steve Grand, along with most true brain modellers, has recognized the importance of chemicals within the brain. He factors 'levels of a punishment chemical' into his particular system, whilst sleep is modelled by 'simply attaching chemoreceptors to their threshold parameters and defining the necessary chemical reactions'. Yet chemicals themselves are not autonomous brain components – it is how they change the configurations of brain cells that counts – and the same chemical can work in different ways in different parts of the brain.

Another approach, albeit still hypothetical, would be to use the nascent technology of nanoscience, which holds the promise of unprecedented control over the structure of matter. In theory, nanoscience might eventually enable us to register the position of every neuron and synapse in the brain, and make a map that could be scanned and copied. But like the human genome project, such an exercise would require no insight into human cognition at all. We would end up therefore with a simulacrum of the brain, not a model that extracted the salient features and left out the rest – rather, nanoscience would give us an artificial brain exactly indistinguishable from a real one.

But for many, the promises of nanoscience depend on the impossible – on changing the fundamental laws of chemistry and how basic bonds form matter. For theoretical as well as practical reasons, then, it is best not to assume that we will soon be building simulacra of biological brains, but to think instead of the more realistic prospect of artificial brains – some kind of model or abstraction of their biological counterparts. A big question is, then, whether such powerful systems will actually rival us in terms of original thought and indeed consciousness – whether they will be our equals on a qualitative level.

Whatever the answer to this question, it seems fairly certain that, on the quantitative level, computers will soon overtake us in brute processing power – so long as we can deal with the problem posed by the physical limitations on the storage of information. And once they do so they will surely be capable of artificial evolution, designing machines more effectively than humans. Such machines will also accelerate other areas of technology. Once computers can use oral communication the web could be superseded by the Machine, a single entity of millions of connected speaking machines. The Machine will know a user's profile of interests and preferences by as early as 2011, and by 2021 will control economic management, so that supply and demand become perfectly balanced. In one sense, the computers will have taken over control of the planet.

But of course an automated system is far from being autocratic, or even autonomous. The prophecies of those like electrical engineer Kevin Warwick, and indeed of futurologist Ian Pearson, suggest an independent-minded executive – a silicon controller, independent to an extent but ultimately answerable to a human – rather than a mere slave-system that makes life easier for us.

'Machines will probably surpass overall human intellectual capability by 2020, and have an emotional feel just like people,' promises Pearson. Well, we have seen that computational powers may well overtake us, but these don't necessarily entail a superior intellect. As physicist Niels Bohr, a Nobel laureate, once admonished a student,

'You're not thinking, you're just being logical.' And as for having emotions 'like people', there is no evidence at all that such soft-hearted mechanical beings could or would come to pass. The critical issue is whether they indeed develop 'self-awareness and consciousness'; despite diverse and competing approaches, we saw earlier that nothing so far in AI or IT suggests that there are any grounds at all for taking such an outcome as an article of faith.

But before we all sink back into complacency, we should examine the alarming prospect that 'relating to machines will be more pleasant than dealing with humans'. True, they may not be conscious, and they may not be hell-bent on world domination – but there is a clear threat that is just as sinister and certainly more likely. First is the matter of how much more positively we shall interact with machines of the future than with our co-species: although at the moment we are used to directing our emotions at a variety of inanimate objects – not just computers but cars and toys as well – in the future we shall spend much more of our time interacting with advanced IT and sophisticated robots as opposed to humans. In fact, we shall inevitably find these artificial interlocutors more predictable, reliable, efficient and tolerant of our temper outbursts, stupidity and egotism. Gradually we could become more petulant, impatient, less able to think through problems, both social and intellectual, and utterly self-obsessed. And the poorer we become at social interactions the more we will seek solace with our cyber-friends. One difficulty for a mid-21st-century family enveloped in cyber-living may be that the family will seem 'boring' to each of its members, compared to the indulgent companionship of the net. And the more people immerse themselves in the net, as some might argue is happening already, so they will cease to develop the appropriate social skills of give and take that Nancy Mitford noted in the mid 20th century: 'The advantage of living in a large family is that early lesson of life's essential unfairness.'

To take things even further, if you never have to consider the thoughts and actions of others – because the cyber-world is endlessly accommodating and forgiving – there might even be a progressive retreat into that world. Generations to come may live each in their own inner world, reacting to and with machines, preferring a virtual time and space and family to their own flesh and blood. We are already sadly familiar with this phenomenon in autistic children; such children have difficulty attributing independent thoughts and beliefs to other people. They do indeed see other people as machines that have no emotions and with whom they are unable to establish a relationship. Could the new technologies be predisposing society to produce a larger number of autistic-seeming children, who would not play a full or active part within what's left of the traditional family unit?

Just as the cyber-friends will help us escape from the old constraints of living in the real world, so too that real world itself will be increasingly dominated by invisible and ubiquitous computing; our grasp of 'reality' and our notion of a stable, consistent world 'out there' might start to disintegrate, as we live from one moment to the next as the passive recipients of wave upon wave of multimedia information flooding into our brains.

But stranger still will be the control those same brains have on the outside world. Our mere thoughts might move objects, and we will witness those around us doing likewise. Although the nature of brain processing probably discounts the possibility that we will be able to hack in to someone else's brain with email via a convenient and trendy-looking implant, psychokinesis – if not full neurotelepathy – might no longer be just the stuff of weirdo-babble. Yet the very *feel* of thinking things to move, of wearing the small devices that enable us to do so, not to mention having internal prostheses that enhance our senses and abilities – all these developments will transform how we think of our bodies, in terms of our abilities and in terms of their boundaries with the outside world. IT and AI will grow into a neuro-technology that blurs reality and fantasy, and dilutes that previously solid 'self' into a wash of carbon-silicon phenomena that will amount to life and living. And if this is *how* we will be, then *what* shall we actually be doing?

4

Work: What will we do with our time?

'So what do you do?' The classic opening gambit almost everywhere in Western society. Rightly or wrongly work defines you. It can give you status, encapsulate in a word your skills and knowledge, and even hint strongly about your predispositions and emotional make-up. Imagine then a society where jobs no longer exist. The notions of 'workers' and 'management' have long been consigned to history, as all the late 20th-century dreams of the human resources industry have come to pass. Instead of a cumbersome crowd of biddable operatives the workforce is comprised of flexible, curious, commercially savvy individuals who are fully aware of their own strengths and weaknesses. These paragons accordingly take responsibility for planning their own career paths. Everyone is an expert or specialist in something, and on the alert for lifelong retraining. On an almost daily basis these multi-module operatives will make instant, plug-in-and-play contributions to small, highly flexible companies with fluctuating daily needs.

Alternatively, the change in work patterns might throw into sharp focus the deep and terrifying question of your intrinsic worth as a person. Just think of the bleak prospect of an insecure, anxious society; most people feel inadequate, unable to keep up with the pace of change or cope with the uncertain nature of their employment. In times to come, far from a simplistic split according to colour of collar, there might be the more invidious distinction of the technological master class versus the – in employment terms – truly useless.

Which is it to be? If the IT revolution is changing the way we view ourselves in relation to the outside world, then its influence on what we actually *do* there will be immediate and far reaching; within the workplace this cyber-upheaval will determine how we interact with other people and things, and hence how we see ourselves. The computer, more than any other single object, will drive the change in work patterns, and even redefine the concept of work itself.

Apparently, it takes some fifty years to optimize a technology – two generations or so for it to assimilate fully, through all the institutions and functions that make up an economy. The computer, icon as it is for a genuine revolution, analogous to the transition from steam to electric power, fits well into this timescale. For the first twenty-five years, say from 1945 to 1970, IT followed electricity in its initial unreliability and inefficiency; it had no measurable impact. Then for the next twenty-five years, into the 1990s, it was still expensive, non-standardized and unreliable, although purchased in large numbers. The workforce did not really know how to use it well, nor management how to apply it.

Only at fifty years old, did the 'new' computer technology become hyperproductive, delivering all its promises simultaneously. At last, at the turn of the century, IT has finally matured into adjectives such as 'cheap' and 'easy to use', with the tsunami of applications and knock-on implications it has for our lives. But just as IT has come of age, so it might be simultaneously doomed – at least in its familiar silicon guise, powered by fossil fuels. As we saw in the last chapter, Moore's Law, which predicts that computer power will double every eighteen months, can't hold true for much longer. The big problem is that the workhorse components of the computer, its transistors and wiring, cannot shrink any smaller than the width of an atom. So, if computers are to be powerful enough to support and sustain the dramatic reality-changing devices that are otherwise technically feasible, then an alternative, fundamentally different type of computational system will soon be needed. The future of work is therefore tightly intertwined with that of the computer, or rather with the issue of what its successors might be, and what they will be able to do.

Up until now there have been four ways of conveying information, by means of numbers, words, sounds and images. But new bio-based technologies, the digitizing of smell, taste and touch along with such slippery phenomena as intuition and imagination, might soon be adding to the richness of our cyber-environment. But biology could make a more fundamental contribution still. Whilst silicon implants might aid the flagging carbon-based brain, so the unique properties of living systems are providing inspiration for utterly different forms of IT: there might even be a case for a novel term, BioIT (biological information technology). Incredible and downright unlikely as such carbon-silicon hybrids might seem, the nascent technology already exists. It might be hard to believe that neurons could communicate readily with artificial systems but they appear to do so with ease. For example, neurons are now being cultivated 'in vitro', literally in glass, along with growth-promoting material, to form narrow tracks in whatever geometry might be desirable for a particular circuit; such bio-wiring could then be hooked up to its silicon counterparts. The neuron would be no more and no less than a new type of electronic component, a 'neurochip'.

Neurochips could indeed become the basic component of the future semiconductor computer. In a normal transistor current flow is modulated by the control of the voltage. Meanwhile, a cornerstone of physiology is that neurons generate a steady-state voltage (the resting potential) that can switch sharply when one cell is signalling to the next with a brief electrical blip (an action potential). Neurons change their voltage in this way by the opening and closing of minuscule tunnels in their membrane walls, through which ions (charged atoms) can flow into or out of the cell. This trafficking of ions amounts to momentary changes in the net charge between the inside and outside of the neuron, namely differences in the voltage across the membrane wall (potential difference).

In a recent ingenious development at the Max Planck Institute in Martinsried, Germany, Peter Fromherz has been exploiting this phenomenon – and the fact that neurons are ultra-efficient at voltage control – using the neuronal membrane wall as a gate contact for a transistor; so far a single neuron can cover sixteen closely packed transistors. To make prototype neurochips, snail neurons are puffed onto silicon chips layered with a kind of glue, placed over the transistors and held in check with tiny polymer picket fences. As the neurons grow to make connections, synapses, so the transistors amplify their tiny voltages.

These snail cells are ideal for the prototype neurochips because they are extra large and therefore easy to manipulate, and they can send electrical signals to each other in the usual way of living systems, as well as communicating with the non-biological, electrical components of the chip. Each cell sits over a Field Effect Transistor, capable of modifying tiny voltages. Although it seems hard to credit that as fragile a biological entity as a neuron could survive in such an artificial and isolated environment, away from the comfort and protection of the cerebral mother-ship, the displaced cells positively flourish. In fact, the scientists developing the project actually have to place physical barriers on the chip to stop unrestricted cell growth, which might otherwise eventually throttle the delicate circuitry that they are trying to establish. So far the team have been working with an array of some twenty cells. They have now devised a chip in which the electrical blip, the action potential, travels to a transistor, and from there to another transistor, and finally on to a second neuron. But Peter Fromherz is currently planning an awesome 15,000 such neuron-transistor sites!

Soon therefore we might see silicon-carbon hybrid neurocomputers with the potential to be far more compact and efficient than the standard machines of today. But some might feel queasy at this blurring of the distinction between living and non-living matter. Could such a system ever be conscious? Would it therefore feel pain, or, even more problematic, develop its own agenda – perhaps that most sensationalist and feared one of world domination? To all these questions the answer would be 'no'. Remember that the biological component, the isolated neuron, is being used solely for its electrochemical efficiency, completely outside of the context of the three-dimensional brain, and indeed away from the brain-in-a-body. Since the system is deprived in this way of all other, as yet unidentified, essentials for consciousness, the risk of its developing any subjective inner state would surely be effectively zero – as it always has been for the vast range of isolated bits of brain found in labs, grown routinely as cells on a dish or slices kept alive for hour after hour in a routine experiment. Instead, consciousness is an emergent property of complex chemical systems within bodies, of which the enormous number of neuronal circuits compressed into brain tissue are still only a part. Here the situation is starkly different: the neuron is merely an electronic bit player, removed from the rich chemical and anatomical landscape of the brain; it is not that much different therefore from the all-silicon systems that featured earlier, just better at the job. Once the silicon-carbon hybrids are in use, it is unlikely that such qualms would even occur to our successors, who will rapidly come to accept the blurring of the current categorical distinction between carbon- and silicon-based systems – just as a century ago humanity accepted horseless carriages.

Meanwhile, at the Technion-Israel Institute of Technology, scientists are using an even more fundamental biological building block, DNA, to tackle the problem of how to make computers smaller: let the circuitry 'grow'. Uri Silvan, the head of the team, sums up the situation:

Conventional microelectronics is pretty much approaching its miniaturization limit . . . If you are really able to fabricate devices with [much smaller] dimensions, you could squeeze roughly 100,000 times more electronics, or even a million times more electronics, into the same volume. That would mean much bigger memories and much faster electronics. In order to do that we need materials with self-assembling properties. We need molecules into which we can encode information which later will make them build themselves into very complex structures. Information is stored in the DNA in this way, information which is used by biological systems to build very complex molecules. We are really trying to copy that idea.

So, Silvan is programming DNA to grow a strand of protein between two electrodes, which is then turned into a wire by depositing atoms of silver on it. An even more radical option still, based once again on DNA, is to dispense with conventional electronics altogether. In 1994 Leonard Adelman, a computer scientist at George Mason University, Virginia, pioneered the concept of a DNA computer. Each molecule of DNA stores information, like a computer chip. And the shift in scale is dramatic: 10 trillion molecules of DNA can fit into a space the size of a child's marble. A few kilos of DNA molecules in about 1,000 litres of liquid would take up just one cubic metre – and store more memory than all the computers ever made.

In a DNA computer the input and output are both strands of DNA, with DNA logic gates (for example, the 'and' gate chemically binds two DNA inputs to form a single output). Although the system resembles a conventional computer, in that they are both digital and therefore can be manipulated as such, the silicon computer has only two positions, 0 and 1, whereas its DNA counterpart has 4 (the nucleic acids that make up the genetic code: adenine, A; cytosine, C; guanine, G; thymine, T). This doubling of the number of positions means that a fledgling DNA computer has already solved, in one week, a problem that would take a standard computer several years, storing over a 100 trillion times more information. This staggering feat is possible because one step in computation will affect a trillion strands of DNA simultaneously, so the system can consider many solutions to a problem in one go. Now add to this appealing arrangement another colossal advantage over silicon computers: since the DNA computer is a biological system excessive heat generation will not be a problem, so it will be a billion times more energy efficient.

The great weakness of DNA computers, however, is that their molecules decay, and therefore they would be no good for long-term storage of information. Another question, raised by the physicist Michio Kaku, is whether they could ever be very versatile, as the solution of each potential problem would require a unique set of chemical reactions. These reservations notwithstanding, DNA computers could be useful eventually for number crunching, within broad classes of problems.

So far, all the biological encroachments that we have been looking at in current computer technology, be they neurochips, neurocomputers, DNA wires or DNA computers, still have a very basic feature in common with the most modest laptop: they are digital – they process information in an all-or-none way. Not so an even more innovative type of computer, first theorized some twenty years ago by the physicist Paul Benioff: the quantum computer. As you might guess from the name, this concept is a machine that works on the principles of quantum physics. Unlike digital machines, a quantum computer could work on millions of computations at once.

If quantum computers really could be developed for practical applications, they would represent an advance comparable to the advent of the transistor in replacing the vacuum tube. In fact, the system could be used initially as a transistor – a means of regulating the current flow, on or off, of a single electron occupying some 20 nm of space (a quantum dot) that thereby generates a 'bit', a 1 or 0. But this most immediate use of quantum transistors is trivial in comparison with the long-term application of quantum theory to computing. The novelty, and power, of this approach lies in the fact that the bit will not be 1 or 0 but could also represent somewhere in between. So, the quantum bit ('qubit') is very different from the bits in conventional computing. Since qubits can be between 1 and 0 (superposition), the potential number of states for the computer is astronomical. A quantum computer with only a paltry 100 qubits, for example, will nevertheless have a mind-boggling 10^{29} (a hundred thousand, trillion, trillion) simultaneous states.

Since quantum systems allow so many possibilities, they can outstrip our fastest conventional chips to the order of some billion times. In practical terms, it means that a system with just forty atoms could track down a single phone number out of every one in the world in just twenty-seven minutes – a modern supercomputer would take a month. Needless to say there are currently severe drawbacks to realizing such formidable computing potential. The biggest problem is that any atom in the quantum computer that collides with another would count as a measurement, and therefore the system has to be completely isolated from the external world. One way round this problem, though hardly very practical, has been to use highly sophisticated and expensive techniques to cool the atoms to near absolute zero, thereby preventing endless random collisions.

A more realistic approach has been to use a technique (nuclear magnetic resonance, NMR) that makes the nuclei of atoms behave by pushing them into alignment with an externally applied magnetic field. Two alternative alignments of nuclear spin, parallel or not parallel to the external field, would correspond to two quantum states with different energies, the qubit.

One objection is that atoms are only so obedient, even when marshalled by NMR, for a few seconds at a time. Even so, the systems built to date still permit some 1,000 operations to be performed at any one time. We have only just begun to develop this new technology, and can hardly yet appreciate its potential. But, as one quantum-computer expert has pointed out: 'All along, ordinary molecules have known how to do a remarkable kind of computation. People were just not asking them the right questions.'

The excitement over future quantum computers is not so much that they will replace silicon systems for word processing or email – in any case our future will probably not include such word-based communications. Instead, quantum systems would be used for large-scale applications, such as cryptography. It is a disquieting thought that no information on the internet would be safe ever again. But then, future generations may be much more habituated to living in the public eye, with privacy as an outdated notion. Protection of individual data might simply be an unplanned consequence of the enormous plethora of data in the cyber-world, so great that no one could be bothered to access it, even if they knew where to start: at the level of domestic lifestyle such activity would be pointless and unappealing. More worrying, however, are the implications not so much for each citizen and their once-private life as for the commerce, politics and security of the nations in which they live: easy code-cracking and infiltration of secret material would have, obviously, devastating financial and military consequences.

How might it feel to work with such powerful machines? We shall see later that a sense of insecurity will probably be the underlying tone of everyday life in the future. Generations to come may well be habituated to a lack of privacy, and take for granted the garnering of information over time frames that today would leave us breathless. Just as we are used to a computer culture that would have been impossible to comprehend in the mid 20th century, so the relatively modest shift in mindset to living with vastly more powerful computers should presumably be easy to make.

By contrast, a basic but possibly catastrophic issue at the heart of the interconnected future of computers and work is the big problem of supplying fuel for those machines. The prospect of life without fuel is not some extremist eco-warrior's prophecy but a very real concern. Our grandchildren may well face up to a crisis where coal is too difficult and too sparse to dig up, nuclear fission considered too dangerous, and solar batteries too expensive; the bleak prospect of an effective return to the Middle Ages, to a life reduced to working, sleeping and eating with only the first two as certain.

Remember that energy cannot be created, only transformed from a limited pool in the universe. And the traditional, easily accessible sources are running out fast. We are currently witnessing an exponential growth of output from Latin America, Asia and in particular

China, which has an increasing energy production capacity of fifteen gigawatts a year. But according to World Bank estimates, developing countries alone will need 5 million megawatts of new generating capacity over the next thirty to forty years, whereas the total world capacity is about 3 million megawatts at the moment. The average citizen of the developed world is using only 1 per cent (approx. 2,500 kilocalories) for staying alive each day, dedicating the remaining 99 per cent (250,000 kilocalories) to making their daily life more enjoyable.

As the need for power is increasing, reserves of fossil fuels – oil, coal and gas – are decreasing. They will be used up between 2020 and 2080, or according to the American Physical Society a few decades later, by 2100. Nuclear energy has a very bad public image – but the only answer for the longer term seems to lie in this direction. Yet even if public opinion could be persuaded that nuclear energy was not as unpalatable as many believe, uranium supplies are also dwindling. What will happen in the very long term? Although there is no simple plan B, the outlook is not completely bleak. Already there is more than the promise of a completely novel way of generating electricity – a different type of battery.

We are all familiar with the batteries in radios, toys and electric toothbrushes, the primary cell that produces energy from chemicals sealed into it at manufacture and which cannot be recharged. Another type of battery is that used in mobile phones, or the lead-acid batteries in cars. These are secondary cells: they are rechargeable. However, there is a third type of battery: the fuel cell, which generates electricity from fuel supplied continuously from outside the cell. The fuel in question, usually hydrogen gas, is passed over one electrode and oxygen is passed over the other. The ensuing flow of electrons is, of course, electricity, whilst the subsequent combination of hydrogen and oxygen yields water (H_2O) as a by-product.

The basic principle of the fuel cell is hardly state-of-the-art high-tech. The first battery of this type was built as long ago as 1839 by a Welsh judge, Sir William Grove. Prototype modern fuel cells already exist, and as such are now in limited use. For example, the space shuttle needed continuous energy, so a fuel cell was eventually developed that ran on hydrogen and oxygen; there were no pollutants and the sole by-product could be drunk by the crew. So, since fuel cells offer a source of energy that does not run down, do not need recharging and produce valuable water, they would seem to be the perfect means of future energy supply. But, as always with new technologies, the components, the batteries, are bulky and expensive, which prohibits their everyday use; yet a still more fundamental requirement is a cheap and easy source of hydrogen, the essential 'fuel' of fuel cells.

Hydrogen has benign environmental characteristics, but it is hard to come by cheaply. It can be extracted from fossil fuels, but that would solve no problems. Another possibility is to split water into hydrogen and oxygen using electrical current (electrolysis) from other renewable energy sources: but what might those sources be? A completely different way of tackling the question, which breaks this vicious cycle, is to turn to biology.

One idea is to extract hydrogen from methanol, which could eventually come from genetically modified trees. And another alternative is proposed by Professor Tasios Melis of the University of California, Berkeley, who is modifying the properties of algae to produce hydrogen from sunlight and water. Melis has made the important discovery that sulphur deprivation dramatically changes the metabolism of the algae so that they can function without generating oxygen. Within twenty-four hours the algae become independent of oxygen (anaerobic), and in so doing activate an enzyme that produces hydrogen in the light. This 'microbial electrochemistry' approach has the appeal of harnessing biological sources rather than energy-inefficient and potentially toxic mechanical ones.

Along different biological lines, Stuart Wilkinson, of the University of South Florida, has provided a complete alternative to electricity with a prototype 'gastro-robot'. As the name suggests, 'Chew Chew' gets his, or arguably her, energy from food – in this case sugar. In Chew Chew's container are E. coli bacteria which metabolize sugar, releasing electrons on one side of a fuel cell; the negatively charged electrons are drawn to oxygen atoms on the other side, hence creating a flow of electricity. The immediate, obvious applications of Chew Chew's technology would be mowers and all farm machinery, which could use vegetation as food. Although there are still non-trivial issues of food location and identification, food gathering, chewing, swallowing, digestion (energy extraction) and waste removal, such a device – like the sulphur-deprived algae – at least holds the promise, given appropriate supplies of food or water and sunlight respectively, of generating electricity indefinitely.

Future generations may well see 'energy' in a different light, if it becomes, one way or another, so closely linked to biology. From our early-21st-century perspective we might imagine that this would make everyone more energy-conscious; but if our successors know nothing else, taking bio-energy for granted might be yet another example of how a mindset is transformed because traditional divisions have broken down – this time between the physics of domestic appliances and living things.

In any event, let's assume that, unconstrained by the laws of physics or energy shortages, IT will continue to shrink whilst becoming ever more powerful; how will it transform the way we work? The first and clearest impact for most of us will be in our physical surroundings. Entering your workplace and logging on to your computer will be very different: passwords will be long gone. Software such as 'FaceIt' already exists that scans some fourteen facial features that do not change. And once in your office you will be surrounded by interactive, smart devices . . .

High-quality visual displays will be able to 'pop out' of small objects such as pens. Philips are already dreaming up ways in which the surface of a desk could 'recognize' tools placed on it. Such 'active' tools would include video-telephony, enabling you to communicate not just with sound but also with body language and gesture. Even the techno-equivalent of a 20th-century handshake might involve the placing of palms on the screen, thereby putting you in virtual contact with your interlocutor. Once-cumbersome monitors and processing units of PCs will be merged into desktops, whilst the currently state-of-the-art Bluetooth radio system for streamlining communications will evolve into a truly single physical system comprising voice- and email, mobile phone, fax, internet access, diary, word processor and video-conferencing facility.

But as the Information Age matures we will not necessarily carry on working in exactly the same way as before only more efficiently. Instead, the new information technologies will change not just how we work but what we actually do – and most conspicuously – where.

Thirty years ago the 'new idea' was the notion of modular office boxes with moveable walls. Another innovation was 'hotel-ing' whereby employees were assigned a phone and workplace on a day-to-day basis, as office nomads. Yet still we cling to the idea of an office, a physical place set aside essentially for talking into dictaphones, into the phone or face to face. Even the current transitory phase, whereby the boss dispenses with a traditional secretary because he or she can email or word process, will eventually fade as voice-interface IT takes hold. A room for just talking does seem then to be increasingly odd. Indeed the basic concept of the office is actually 150 years old, and may be about to become as obsolete as outdoor privies are today.

Take, for example, IBM in Chicago. The company used to employ 10,000 on site; now the workforce is down to 3,500, of which 80 per cent work from home, and the building itself is up for sale. Similarly, Motorola is planning to relocate 40 per cent of its workforce in their homes. One general prediction is that soon a third of the workforce will be doing gainful work from home. There will be a rise in virtual organizations operating more flexible, demand-oriented production networks. But as more people work from home the 'beehive' mentality of humans will surface as an increasingly important factor – the perhaps obvious human need to feel a valued part of a busy, thriving community, which is not met by living as an isolated hermit with no immediate incentives or constraints on performance. Perhaps, in the future, that very need might be muffled and muted by the habits of a life dominated by technology, or alternatively the technology might itself be so good that the need is met artificially; but until that transformation – either in human nature or in IT – is complete there will be an uneasy tension between the obvious advantages of working at home and the feelings of depersonalization it could bring.

In the meantime, not just the place and manner of work but also the type of work available is inevitably already changing as a result of the cyber-revolution. Michael Dell, when starting up his now hugely successful PC company, made the then daring decision to offer his product exclusively over the internet. In one simple step, the burgeoning internet economy is reducing the costs of transaction and of bringing a product to market. The old rationale for forming large companies – namely to reduce the costs of gathering in materials and economizing on other physical desiderata – is no longer sacrosanct; accordingly, we shall start to see an ever-increasing proliferation of alliances. The trend can only continue, therefore, towards a breakdown in monolithic organizations in favour of smaller, more virtual units that, although independent, network with each other.

Of course, some companies will be larger than others, but the watchword in the knowledge economy will be 'specialization'. Instead of battling other wannabe leviathans in a winner-takes-all competition, each organization will focus on doing only what it is best at, outsourcing the rest and networking synergistically with other small companies. Imagine a galaxy of small enterprises, with subcontracts, as satellites around a bigger company. The ideal, almost a caricature, company of the future would be all knowledge and no assets at all, except webs of flexible relations with suppliers and subcontractors.

How will this change the thinking of the workforce of the future? One immediate result will be less competition, and therefore a cooperative culture, yet with less security, involving a much more frequent change of jobs. In fact, the concept of a 'job' as we know it may well disappear altogether. Driven by the just-in-time agenda on which small, high-risk businesses thrive, firms will perhaps bid for employee time almost on a day-to-day basis.

Most certain of all is the prospect that this decade will deliver the final death-blow to the expectation, already on its last gasp in the late 20th century, of a job for life. We shall have to come to terms with more down-time unemployed or retraining, and will do so by taking 'portfolio' careers into our own hands, with little regard for corporate loyalty. This death of the 'job' concept and rise of the portfolio-toting freelancer is by no means limited to white-collar workers. But we should not underestimate the shift in attitudes that will be required for such changes to become universal. There will be a big shift in requirements towards personal and communication skills, rather than just intelligence. Management expert Tom Peters says that workers 'must no longer count on corporate hierarchy for their careers or their identities', but instead 'must act like freelancers and entrepreneurs, building a portfolio of skills and accomplishments they can use to negotiate the next job'.

The old structure of workers and bosses has been eroding for decades. This erosion will soon be even more widespread. In 1970 it took some 108 men 5 days to unload a ship of a particular size. However, since containerization, a comparable task is now achieved by 8 men in a single day, amounting to a reduction in man-days of 98.5 per cent. The development of enterprise software is currently doing for the white-collar workforce what containerization, forklifts and robots did for blue-collar workers: the managerial sector may be in for a staggering 90 per cent reduction in the next ten years. With the loss of the low-skill, repetitive jobs that will be obsolete in the robotic age, and the rise of newly proactive other ranks, middle management will go. Not only will the rise in dotcoms offer the opportunity to cut out the middleman or -woman but also the trend for outsourcing will cut labour costs and the attendant administration.

E-commerce itself is ballooning as an industry – netting some $300 billion in revenues in 2001. IT has now grown to the size of the American auto-industry; almost half the workers in industry either produce or are now intensive users of IT. As well as causing massive reorganization in the workplace, the IT revolution has inevitably recast the nature of skills required. As the US Department of Labor has indicated, you need only to look at the job ads to see how rapidly the job market has changed even within the last ten years. Only a decade or so ago there was still a demand for typists, repair technicians and switchboard operators. Now we search for webmasters and desktop publishers.

Even within the professions, e-consultations and cyber-surgery may eventually oust flesh-and-blood lawyers and doctors; gradually these expensively and exhaustively trained individuals, imperfect repositories of knowledge, will give way to constantly updated robotic and computer systems that are far less fallible and can cater for every contingency. And outside of the professions one possibility is that people could become extensions of computer-driven production, a more high-tech and interactive dehumanization than the assembly line. Such operatives might become, like Silas Marner, distanced from the natural world, methodically working at a task which offers no personal satisfaction but demands rather 'the bent, tread-mill attitude of the weaver'.

Then again, job descriptions could become so flexible as to be meaningless. The age of just-in-time operatives, geared to meet the needs of just-in-time production, will be upon us. For most of the next generation, flexibility in learning new skills and adapting to change will be the major requirement as they work their way through smaller companies – 99 per cent of UK businesses in 2001 employed less than fifty people. Once again, the trend continues away from the notion of the 'job' towards setting your own agenda: you'll then get on with doing whatever needs to be done, in whatever way seems best, with whatever skills you have – and those you don't have you'll outsource.

However, there will be some fallout from this happy scheme of things; probably a significant sector of the population will be simply unable to metamorphose into a perpetual-learning mindset. Most likely, a critical factor will be the degree of stress involved – you could be frantically scrabbling for new skills, anxious that, any day now, your current expertise will be deemed obsolete. Who knows if we would wish to conduct our working lives in such a state of red-alert, or whether we would find the constant change stimulating. In any case, we might not actually be up to it as our brains age, regardless of the era in which we are born.

Up until now, at least, older people have found learning new things difficult. Not only are the molecular mechanisms less sprightly but at the more macro, cognitive level too the 'fluid' intelligence of the young brain, mopping up information like a sponge, gives way to the 'crystalline' intelligence of the mature brain, evaluating any incoming information against what is already known. Only a fraction of the incoming traffic ceaselessly bombarding the maturing brain will therefore finally be assimilated and itself add to the sales-resistance against any subsequent novelty. And an age-sensitive discrepancy is exacerbated by a one-off aspect of early-21st-century life: there are those for whom the net and web have always been part of their lives, who are infinitely more at home with keyboard, screen and mouse than with pen, paper and book – and on the other side of the divide, those for whom the reverse is true.

It's easy to imagine being endlessly on the back foot, trying to keep up with the demands of an anonymous screen-based boss, and over the longer term always being just-out-of-date rather than just-in-time with the skills that you have. And lurking and smirking in the background would be the minority of elite young technocrats who know everything and are conspicuously in control of their lives, writing their own job descriptions as they go along. The average employee might feel demoralized, inadequate and old as they compete for niches alongside a younger generation so comfortable with change and shifting realities, with interactive cyber-worlds of no permanence, that they ceaselessly adapt to new demands – a generation for whom technology is as natural a phenomenon as life itself. Meanwhile older and/or non-technically adept workers might be unprepared, untrained and frightened by the prospect of taking their careers into their own hands.

Yet there is a more positive possibility, even for those of us on the wrong side of the age-divide during this time of transition. It need not be all doom and gloom for those of us who can remember the Cold War and the advent of TV and of avocado pears. Since commerce will become more global, there will be a need for local perspectives and hence for the leaders who can make a difference in the virtual workspace, whilst being at the same time more mobile and international. Some actually think that baby boomers, far from being risible relics of a bygone era, will constitute this wiser, senior stratum of the workforce within the next few decades, staying on at work and offsetting the decline in birth rate in first world countries.

But sooner or later the majority of the grey workforce will be in crisis: the next few decades of this century will surely witness a gaping fissure between the work-style abilities of the baby boomers (born between 1946 and 1965), generation X (born between 1966 and 1982) and generation Y (those born after 1983). It could well be an irony of the first half of the 21st century that just as the older workforce is increasingly unable to meet the new demands, so more work becomes available for senior citizens.

Although we tend to imagine that, as the computer gradually takes over the grunt-work of work, we will automatically have more leisure each day and indeed over our lifespans, the increasingly pervasive IT might, paradoxically, encourage people to stay at work beyond the traditional, 20th-century Rubicon of retirement age. After all, IT and a service-oriented economy will offer more flexibility and gentler hours, along with growing opportunities to work from home, with consequently fewer demands on physical strength. At the same time, older people could well be fitter and in search of brain stimulation and a sense of self-worth, not to mention income. We might find that there is no longer such a clear schism between work and retirement – a comforting thought bearing in mind that within the next ten years ninety million of the population of Europe will be over the age of sixty.

Some even think that the concept of retirement will disappear. Older people, and the society in which they live, will be ever less likely to accept the fate, played out nowadays in so many old people's homes, of vegetating in a chair, staring at the walls. At the moment it is difficult enough, in terms of resources, to ensure that the residents are warm, well fed and clean. But in the future it is very likely that, at very little cost, we will be able to devise some means of stimulating the aged brain. As computing decreases in price and becomes ever more ubiquitous, invisible, interactive and smart, so older people may have the opportunity to take themselves through a series of interactive exercises, interfaced in a way that fits their physical capabilities and needs no external nursing supervision: a kind of brain gym. Perhaps, once everyone is IT-instinctive rather than merely IT-literate, even the 'natural' tendency for the ageing brain to slow down might come into question. Improved healthcare could ensure that we will have not just 150 years of wisdom, but enough energy to use it and to extend our productivity.

But although a grey workforce of lively minded and able-bodied senior citizens seems as plausible as it is desirable, it is unlikely that the elderly will have the absolute physical prowess of those in their twenties. Moreover, they will be contributing a different type of intelligence, crystalline wisdom rather than the rapid adaptation skills that characterize the younger brain. Such contributions could have a range of outcomes. First, the outlook of the senior workforce may jar with that of younger colleagues; or secondly, they may enrich the workplace and product; thirdly, perhaps age differences will not mean differences in performance since people will all have been exposed to a homogeneous cyber-world for most of their daily lives, and will thus be much more homogeneous in outlook than different generations are nowadays. If physical health is no longer a means of discrimination, and if knowledge is to be stored with easily accessible cyber-experts, and if much of life in general will be virtual, then for the first time we might be squaring up to a society with a far less clear-cut demarcation of distinct generations. Society may end up the poorer for lacking the diversity of different groups with very different outlooks on life, and diverse tastes in foods, fashions, music and so on. On the other hand, a culturally uniform society may be a small price to pay for a healthier and more fulfilled populace – so long as the individual really is more fulfilled.

In any event, by 2050, the workforce as a whole will be very different: only about half of the US population will be non-Hispanic whites. So, we can expect not only an older but also a more multi-cultural workforce. There will also, most likely, be even more women. In the USA already more women are finishing college than men. Moreover, the traditionally subordinate sex appears to have real leadership potential: currently over 50 per cent of editors and reporters, and 54 per cent of authors, are women. Women (70 per cent) are enrolling in college at a higher rate than men (64 per cent) too. But probably these increasing numbers will flatten out at some stage soon, once issues and factors relating to childbirth and child-rearing take effect – if one assumes that the nuclear family, and a woman's role in it, remains as it has been – and immigration and welfare laws are embedded.

Indeed, the phenomenon of the strained and burnt-out superwomen who strive in vain to 'have it all' is now contrasted by movements advocating a complete reversal back to the 'surrendered wife', captured so well, at a time when such an idea was ridiculous, in the cult 1975 film *The Stepford Wives*. The stay-at-home wife is financially dependent on her husband, taking on cooking, cleaning and childcare whilst the husband is out of the home working for most of the day. The good news – in fact the argument for 'surrendering' – is that women are allowed not to strive to have it all, and to spend much more time with their children. However, the bad news is that this advice extends to allowing one's husband to win arguments, and to concede to him on all family matters, including financial ones. It is hard to see how maintaining such a mindset in the late 20th century, let alone in the 21st, wouldn't place a strain on the women trying to unpick several decades of the culture and education in which they have grown up. On a more prosaic level, in a world where the home will be a very different place from what it is now, where it will merge so seamlessly with the workplace, the role that the 'traditional' wife might have is far from obvious.

John Gray, author of the acclaimed *Men are from Mars, Women are from Venus*, suggests that men used to have the edge at work because they could devote more time to it; but now the flexibility of the computer and the trend towards increased home working, coupled with the natural communication skills of women, all amount to a shift in the gender balance in senior positions. But a more fundamental consideration is that in a future in which, it seems so far, the very nature of individuality will be threatened, and in which we will become the passive recipients of incoming information and immediate sensations, clearly defined roles within a home and family will no longer be appropriate or helpful. We saw earlier that the dynamic of the family unit will be shifting, not just as a result of serial monogamy, divorce and remarriage but also due to the growth of gay and lesbian unions, and a possible blurring of gender roles.

Nonetheless, Christopher Jones, Professor of Political Science at Eastern Oregon State College, sees the prospects of this new century as still far from perfect for women. In legislation and in the boardroom there is still gaping under-representation, a 'glass-ceiling' syndrome and institutionalized sexism. Yet over the last thirty years the time a married woman with children spends outside the home has doubled. Parents in the USA spend fewer than twenty-two hours per week with their families, whilst at least one in five households are responsible for informal care of a friend or relative over fifty. This figure will more than double in the next five years. A large number of people therefore may soon find themselves sandwiched between caring for the very old and the very young.

A hypothetical working day in 2030 has been predicted to run as follows: get up very early, to deal with domestic errands, then spend the next few hours in virtual meetings amidst 3D images of your colleagues. A video camera will transmit back to your supervisor a real-time record of you working to your maximum productivity. You may then spend an hour or so with grandma before picking up the children, putting in some alpha parent time, and recouping lost hours by working from 8–10.30 p.m. In essence, your workday will be flexible, disconnected, personalized – but lengthy.

So, more even than for today's workforce, frustration and mental fatigue could be part of the working day. 'Desk rage' has apparently already hit the workplace, a phenomenon characterized, as might be imagined, by stressed-out employees indulging in rude and inappropriate behaviour. However, the increased flexibility needed in our complex portfolio lives will lead to a blurring of the line between work and private life: we will, increasingly, have fewer qualms about sorting out our electronic domestic jobs, like organizing our home arrangements and online shopping, from the office whilst at the same time using the enhanced IT opportunities to do more and more work from home.

And, like the merging of work and retirement, the merging of work and home together with a lack of structure and security in the workplace could all lead to still more misery. In his book *Britain on the Couch*, the psychologist Oliver James examines why the British are more depressed now compared to over half a century ago, when they were materially far worse off. He concludes that unrealistic expectations, increased competition, unstable relationships and the strain of being 'individual' have all induced a national lowering of the neurotransmitter serotonin.

Serotonin is a pervasive, fountain-like chemical messenger in the brain. When serotonin is low, depression ensues; indeed, Prozac works primarily by elevating brain serotonin levels. We do not know why a low or high level of this particular chemical changes the configuration of working groups of neurons, which in turn translates respectively into negative or positive feelings, but we do know that such fluctuating levels can be modified by the environment, for example, by your perceived status in society. That said, it is important to remember that we can't just jump from a chemical transmitter, like serotonin, to a sophisticated mental function, like fretting about status. Rather, the chemical is a bit player contributing to a change in the configuration of neuronal networks that in some as yet unknown way matches up with changes in mood. The notion of a 'low-serotonin society' is therefore shorthand, describing a collective, general mental dysfunction, an index of which might be, among multiple other indices, lower levels of serotonin.

In any event, the surprising frequency of the gloomy mindset in our comfortable modern lifestyle corresponds also with the conclusions of psychologist Mihaly Csikszentmihalyi. Csikszentmihalyi has spent much of his life engaged in a longitudinal study of human happiness, from which he has developed the idea of 'flow'. Csikszentmihalyi has found that the level of happiness is not related to per capita income. Rather, we apparently each have a kind of internal happiness-thermostat, irrespective of our circumstances. Lottery winners, for instance, are down to their pre-win 'happiness level' within sixteen months. But our levels may rise or fall in different environments: executives are happiest at work whereas clerical workers are happiest at home, and assembly-line workers are different again, being happiest at bars and in cinemas. Happiness is not a function of what you own but, once again, how you perceive yourself in relation to the society around you.

The impact of the IT revolution on work in the future could have either of two effects on our collective mood. Perhaps the increasingly passive role that later generations will adopt, together with a loss of professional identity, will lead to a fall in self-esteem, or even the loss of a sense of self altogether. If society becomes one which places emphasis on achievement and competition, on going it alone, braving the insecurity of new and unstructured workplaces and learning new skills, then the serotonin levels in the brains of many look set to plummet. This unfortunate outcome will be even more likely if, para-doxically, the IT revolution that has triggered these changes at the same time robs us of the ability to be proactive, to think without computer aid, to have a clear sense of identity compared to the outside world and to others – indeed to have 'normal' relationships. Without a clear sense of identity, with no motivating need to feel fulfilled, sedated by easy cyber-experiences – why make the effort?

The much rosier alternative would be that the new technology will enable the self to be more readily expressed. Imagine therefore a multifaceted workforce, ranging from the quick-witted and quick-learning youth to the clear-minded and wiser elderly, a workforce unencumbered by bigoted cultural baggage and unconcerned, thanks to flexible, distance-work patterns, with gender or childbirth. Each individual would be aware of their own skills and preferences and be prepared to develop their own career path without comparison with anyone else. Alert to shifting demands, they would want to find things out and to help improve the world; above all – contrary to human nature as this might seem – they would actually enjoy working.

Both these scenarios are obviously overly simplistic caricatures. Most likely there will be a spectrum between the two, but the big question is how polarized that spectrum will become, how far along, and in which direction, the majority of humanity will be positioned. The positive outcome places great hope on the bulk of society being far more adept at high-tech, confident, curious and self-sufficient than it has been in the past. The alternative would be a truly divided society with those who are not intellectually agile or outgoing increasingly at the mercy of the minority who are. Depending on energy supplies, economic growth and the degree of automation, the disenfranchised majority might be engaged in highly routine tasks, 21st-century equivalents of Silas Marner; alternatively, they could be condemned to a life emptied of purpose and filled with leisure. But unadulterated free time does not appear so far to have caught on.

The great visionary George Bernard Shaw predicted that in the future we would need to work for only two hours a day. Similarly, the economist John Maynard Keynes prophesied back in the 1930s that the working week a century hence would be a gentle fifteen hours. Even into the 1960s the main prediction people were making about work was simply that there would be less of it. One happy picture was painted by *The New York Times*: 'By the year 2000, people will work no more than four days a week ... in an annual working period of 147 days [on] and 218 days off'. And similarly, in 1966, from a mandarin at General Motors: 'People will start to go to work at about age twenty-five. Six-month vacations would not be out of the question.'

Even back then, however, it was clear that lives unfettered from the routine of work might take some rethinking. In 1966, an article in *Time* read: 'By 2000, the machines will be producing so much that everyone in the US will, in effect, be independently wealthy. How to use leisure meaningfully will be a major problem.' In the same vein some two years later, Arthur C. Clarke, author of the work that inspired the film *2001*, wrote on the theme of the real 2001: 'Our descendants [will be] faced with a future of utter boredom, where the main problem in life is deciding which of the several hundred TV channels to select.'

For the first time ever, we human beings may be confronted with the mind-boggling luxury of leisure on a grand scale – not just snatched vacations or short-break weekends, but day after day freed from hunger, cold and pain. The question is far more difficult than it might at first seem: it is not as if suddenly robots will liberate us from the Dickensian factory floor letting us loose into the nirvana of the suburban back-garden *circa* 1950. Rather, we have to ask ourselves those adolescent questions about what our lives actually mean, and, once the demands of survival are removed, what we are going to do with our time.

But strangely, it seems that time is not as high on the agenda as better living conditions, more exciting work, more stimulating recreation and more freedom. A sense of fulfilment can be gained through work as much as, or perhaps even more than, through leisure activities. Moreover, being busy carries high status. How often do professional folk, on meeting, exchange exasperated sighs of fatigue and helpless head-shakes at the dizziness of their schedules? Yet rarely does anyone (me included) confess to taking any counter-measures. The faster that new emails push those of a few minutes earlier up and off the screen, the more your cell phone sings out its tinny call sign, the more important you must be.

This premium we place on status might explain the current tendency to exchange leisure for a higher standard of living, even though we have less time to enjoy it. Just reflect for a moment on the amazing statistic that if productivity gains were translated directly and completely into free time, we would have to work only half the year – but would have a standard of living closer to 1949 than 1998. Until the 1940s unions negotiated for shorter working hours; now their biggest concerns are salary and security.

Some voices in the wilderness may be starting to plead for us to stop devaluing free time, claiming that work is a means to an end not an end in itself. But is this maxim really robust for the years ahead? After all, if it is human nature to need status, a sense of self and self-fulfilment, then work will meet those needs more obviously and easily than leisure. Perhaps leisure time, because it is down to you to spend as you wish, sneakily forces you to question what you are and what your life is all about. In the old days there were hobbies. The labouring classes had clearly demarcated slabs of time away from the coalface or assembly line: traditionally, racing pigeons, allotments and sport led to a sense of identity, skill, fulfilment and pride that would not otherwise have been generated from monotonous and hard daily grind, which effectively made one operative the same as the next. But now, given a world of physical ease and sensual gratification, of freedom from drudgery and pain, a cyber-world of instant information and fantasy, what will you do with the nooks and crannies of free time opening up within your life-narrative into a lengthy old age?

One 'solution' will be to blur the distinction between work and play, and by blending them much more preserve the all-important element of status. According to findings from Penn State University, free time in the USA is up from thirty-five hours a week in 1965 to forty. But we tend to underestimate our periods of leisure since free time is now so fragmented: some twenty-five hours is taken during the working week, from Monday to Friday. Meanwhile, the weekend has 'simply become an extension of the week', concludes John Robinson, co-author of the 'Americans' Use of Time' study. Increasingly our lives, and those of our children, are programmed – so leisure just doesn't feel like leisure. Surveys show that already we feel less and less free, and more pressed for time. Cell phones and laptops extend the working day and increase the workload, rather than compressing it. Leisure is so structured and constrained that it is no longer the relaxing alternative to work – for example, a gym in the USA now opens at 4.30 a.m. to accommodate those who want to work hard *and* keep fit, and thus have to cut down on that most indolent and sensual of activities, sleep. Increasingly we are storing up leisure time for old age but, as we have just seen, the future seems to hold the promise, or threat, of an old age that is a far cry from the 'pipe and slippers' retirement of the mid 20th century.

In the previous century, for most of society, work enabled you to eat whilst you had fun and found fulfilment in activities such as going down the pub, playing sport or going to the cinema. Now the ideal workplace offers both fun and fulfilment just as readily as the home offers a place to work. A reduced working week of thirty-two hours may well increase leisure time to sixty-four hours, but we are no longer private; unless you take extreme and unusual measures you remain available 24/7. 'The cell phone gives me the freedom to play golf, but it also means that I can be interrupted as well. It's a necessary evil of an ever busier world we live in,' explains one 53-year-old car dealer.

The blurred line between work and leisure, then, is both positive and negative. Yet the habit is embedded in our culture to fill time, with work or leisure – we must be emphatically *doing something*, accruing status or seeking oblivion as a sensory sponge. Professor Robert Levine writes in *A Geography of Time*: 'The idea of doing nothing is so foreign to us. Clearly, in the US, doing nothing is bad. People sitting and staring – there's just something wrong with that.'

The problem that we have with sitting and doing nothing at all, not even indulging in a hobby, might be because such vegetation holds no status or sense of achievement at all. We already prefer, it seems, an endless chain of homogeneous waking hours where work and play merge, and we can perhaps sublimate the frightening question of who or what we are that would otherwise be posed by stretches of unstructured time. Until recently, the brute needs of survival constrained how long we could spend in pipe-and-slippers mode in any case. And we have made our priorities, so far, such that truly free time is minimized. But what if the automated cyber-future denies us our work, and offers us no alternative to leisure?

One possibility is that we will choose to 'go real', like the 'real room' in the future home in Chapter 2. Because we will be so unused to low-tech personal interactions such as having a gossip face to face, or going for a walk together, a visit to a 'family interactive sports and entertainment centre' might be an attractive novelty. At Mount Florida in Broward County, like CenterParcs in the UK, there is an area enclosed under a dome – in this case some 800,000 square feet. Here the 'leisure-maker' can ski, skate, toboggan, hang-glide, rock climb and body surf, as well as indulging in a movie, shopping, eating, walking through an aquarium and so on.

The innovative feature at Mount Florida, however, is a personal video-camera strapped to your head, which enables you to relive your experience and to keep a record of it, and your personal profile, at the centre. The intriguing question, though, is this: which experience would be the most valued? On the face of it, the video would be little more than the beach snapshots of yesteryear or today's home video. But a worrying feature is the extent to which you are living on film.

I once went to a wedding where, after the ceremony, we were all left in the hot-oven air outside the church whilst the bride and groom went through the whole ritual again – just so that a benighted relative could capture them in close-up on video. It seemed to me, as I stood in the unforgiving glare of the noon sun fantasizing about a cold, fizzing glass of champagne, that the marriage ceremony had really taken place in order to make the video, as opposed to the other way around. As the sweat trickled down my back the bizarre thought took hold that it was the record that was the all-important goal, not the wedding experience itself.

Could it be, then, that in the future we will be so used to living life through the screen, so used to replays and freeze-frames, to editing and air brushing, to mingling the virtual and the real, to experiencing heightened colour and sound and even smell, as well as explanatory commentary, that the moment-to-moment activity in the real world will pale, literally, by comparison? An alternative, but equal, deterrent to 'real' activity could be that we become so unused to the physical interaction of our bodies with the outside world that the whole experience would be rather frightening and too intimate – a little as killing, skinning and butchering a rabbit for the pot was second nature to our ancestors but would be 'too much' for many people today. Direct stimulation of the senses – wine, women and song – has always offered a way of spending precious free time. But the new technologies are changing the degree to which we wish to be stimulated, and offer alternative cyber-stimulation.

And as well as second-hand stimulation there is the prospect of a completely second-hand life. Watching other people, fictional or real, is, of course, nothing new. Our favourite activity is currently watching TV, for an average of four hours a day, although a trip to the cinema still holds entertainment value – the whole experience of going to a film contributes to a more complete escape from reality. Julie Taymor, director of the film *Frida*, draws an analogy between watching a video at home, compared to a visit to the cinema, and the experience of praying at home, compared to going to a place of worship. She argues that people like ritual and awe, structure and comfort in the face of chaos. And the home-bound internet doesn't awe in the same way as the cinema, just as the immediacy of the human voice, laden with innuendo, has so much more impact than email. The world wide web is 'too safe, too anonymous and too antiseptic'; we are attracted to the big screen because: 'We want to be touched emotionally, be viscerally moved, perhaps have our minds challenged, or at best blown. We travel to a different place when we enter the world of the storyteller. Some call it escape; some call it experience.'

Live events, or even being part of a cinema audience, heighten sensations but technology may change all that, as it will our expectations, habits and needs; in the future, such intense stimulation could be beamed from the walls of your own home. Take watching sport as a particularly good example. Mark Leyner, a novelist and screenwriter from New Jersey, USA, reflects on the endless need to spectate: sport especially offers one of the last places where the improvisational and the unexpected occur. However, he regards as 'relics of the past' the 'self-aggrandizing preening of the ringside celebrity and the self-annulling ecstasy of the anonymous face-painted fan' – in short, simply being there. Already some of those attending sports events, most usually in the USA, watch Jumbotron screens, suggesting that 'being there' is not the top priority nor indeed the sole reason for watching. Leyner thinks, contrary to the view that we like being in a seething mass of sweaty humanity, that there will be no valid reason to attend live sports events 'except for the opportunity to begrudgingly share cheese-drenched nachos with complete strangers or stand in line and chat with other people who also have to urinate badly'.

Rather, the appeal of cyber-intimacy and the option to control sensory intensity will outweigh the first-hand experience. There will probably be an increase in the interfacing of real, physical sports and the cyber-world: the next generation could have the opportunity to compete with top-level athletes on computer games. At the same time there are already a surprising number of spectators of networked computer games, and this trend can only continue to grow. Changes in work practices, the dominance of IT and the rise of robots will also reduce our personal risk, and perhaps therefore place emphasis not on feelings, raw sensations as such, but on thoughts; computer-enhanced

TV will zoom in on your particular curiosities and predilections, even as personal and bizarre as an edit that highlights the times when 'players groom themselves and spit', to take Leyner's example. Data-mining will pervade all aspects of our lives, even our leisure and pleasure; as a result, the appeal of a world that caters for your desires and is under your control, although second hand, will be greater than that of the first-hand press of a real world that won't comply with your personal dictates. We can take this idea to extremes, and hazard that in the future each of us will have our own mix of programmes tailored to our interests and proclivities. 'Real' behaviour will be seen as too messy, antisocial, unhealthy and time-wasting. Instead, all leisure could be conducted on the couch.

Of course quite the opposite may occur, as a reaction to the cyber-world, and very dangerous activities may become popular. For example, as well as the established means of risking life and limb in air, water, mountains and caves, by white-water rafting, jumping from planes and so on, futurologist Ian Pearson predicts that we will see a rise in new high-tech, high-risk sports such as 'zorbing', whereby you roll down a hill anchored in an inflatable sphere. However, the future of these sports may well depend on wider lifestyle factors.

Usually, leisure activity offers a counter-balance to what we have been doing most frequently and most recently. If we have been sedentary, working or studying alone, sleeping well and perhaps leading a monotonous, routine life for a few days, then the appeal of stimulation is obvious – be it white-water rafting, dancing or playing sport, or at least watching it. On the other hand, the excessive stress, constant vigilance and rapid decision-making that characterize many modern jobs tend to steer us towards 'chilling out', reducing stimulation in the immediate environment and slowing down the pace of living. For example, a Stress Reduction Center in New Jersey, USA, offers 'a little bubble of calm in the craziness of a mall'. For $10 you can enter a tinted-glass booth and sink into an ergonomic lounge chair. A virtual bombardment then follows – the sights and sounds of a tropical rain forest, which is supposed to induce brain-wave activity conducive to pleasant, dream-like visions, as time-out from the chaos of the street.

This centre is the latest in a long line of environments, albeit particularly small and synthetic in this case, that offer a way of manipulating your mindset by a wholesale manipulation of the input to your brain. The same net feeling of overall relaxation, however, could be just as readily induced by being on a mountain. I have long been intrigued as to what net brain state generates that final feeling, common to both situations. Similarly, what common brain state generates the overall feeling of 'joy' that can be induced by such otherwise disparate phenomena as downhill-skiing, orgasm, fast dancing, or bungee-jumping?

If we knew more about the configuration of working neuronal assemblies that click into a particular pattern across the brain corresponding to such net common feelings, although derived from different routes, then clearly we could understand why, for example, certain drugs such as Prozac work as they do. Prozac manipulates the very same transmitter system, serotonin, that is rapidly changed when your perceived status changes in a group. To reiterate an important point, however: the molecules of serotonin do not have high status or happiness locked away inside them. The big question is how drugs such as Prozac transform the global brain landscape, and how that landscape translates into an emotion.

Drugs change emotions directly. Already the burgeoning drug culture is offering a route away from one's problems into a temporary oblivion. As I write this, the UK is teetering on the edge of decriminalizing cannabis, perhaps even legitimizing it completely. Instead of coming to terms with, if not circumventing, the difficulties of life, we increasingly seek the chemical equivalent of shutting our eyes and putting our hands over our ears; surely we will soon be living in a society of under-fulfilled, glassy-eyed zombies ripe for control by a minority, like the consumers of the happiness drug 'soma' in Huxley's *Brave New World*.

Drugs are a sledgehammer means of reconfiguring working brain connections more dramatically than the far more subtle manipulation of the environment through the experiences of daily life. The use of drugs might therefore offer for many a more direct and efficient route to bring about a counter-balance to their work, or provide a substitute in its absence, by offering excitement or relaxation. Everyone knows that different drugs can achieve either of these ends: alcohol, cannabis, benzodiazepines, morphine and its derivative heroin are all depressants of the activity of the central nervous system. Amphetamine, ecstasy, caffeine and cocaine have an opposite, stimulant effect.

Note that in the list above I have purposefully interleaved drugs that are both prescribed and proscribed, and indeed which vary enormously in their potency and potential danger, in order to focus instead on the type of effect that they have on the brain. Although all these drugs work to different extents and their precise biochemical action varies, their action on the brain results in either of two types of pattern across the brain. We have yet to discover what the actual patterns of connections are that relate to stimulation and relaxation respectively, but once we do know – perhaps thanks to advanced imaging techniques – then it may be possible to drive the brain into that precise state with certain types of direct brain stimulation, or even using highly specific external software comprised of appropriate rhythms, shapes and colours. Although it would probably not be feasible to implant a memory or idea into the brain, it might eventually be possible to identify the more basic and generic configurations that underscore a particular human emotion – the final common pattern of connections driven by a smile, a drug, a song, a dive or a jump. Feelings could then be manipulated directly, and, most significantly, remotely – not just by you but by someone else.

But even in *Brave New World* the citizens worked. What about the real future? We have seen that IT is only now coming of age, and bringing with it completely different systems of computation, in the form of silicon-carbon hybrids as well as quantum computers, which will challenge any secrecy or privacy that we thought we might still have had. Accordingly, IT will transform the workplace, the workforce and the manner of work. One scenario, as we have seen, would be that the vast majority feel so anxious about keeping up with the just-in-time-skills mentality that they seek solace in direct chemical comfort or some other activity, say the second-hand narrative of a fictional or real screen-based hero, that distracts from a sense of inadequacy.

Alternatively – as Huxley in the mid 20th century could never have foreseen – sophisticated and pervasive automation could free from work the many who are not highly trained technocrats; but the long hours with nothing specific to do could pass very slowly, unless one had a goal, a sense of purpose. Again, powerful stimulation of the senses, either with drugs or with IT, might offer some distraction, but not the sense of status and fulfilment that seems so important to the human condition.

Finally, the most positive scenario of the satisfied worker, happily retraining and managing a complex career portfolio, is one where work and leisure merge; but even here doubts about motivation might creep in. If the majority can live life without work, then why not you? So you enjoy it – why? In a world that offers every physical comfort, what exactly are you, oddball that you are, now trying to achieve?

Adam Curtis, the TV producer who made the ground-breaking series *The Century of the Self*, maintains that it was only in the 20th century that we saw the creation of an all-consuming 'self'. Freud's work on human needs and drives unwittingly led to the belief that the pursuit of satisfaction and happiness was man's ultimate goal. His nephew Edward Bernays then exploited Freud's theories to coax people into wanting something that they did not need: for example, women were persuaded to smoke by associating lighting a cigarette with lighting a tiny 'torch of freedom', hence any woman who smoked was a free-spirited individual.

By establishing the notion of the self, an 'ego', it was possible to suppress deep destructive desires – Freud's 'id'. In less humanitarian times, this id had been manipulated directly with the 'soma' of the time – 'bread and circuses' in the case of ancient Rome – or channelled into tribal rituals or suppressed with harsh and swift punishment. But in a democratic and commercial modern society such courses of action were no longer options. Instead, by cultivating the concept of a self, Bernays and others have been able to cultivate a means of seeming to give people what they think they want – an indulgence and celebration of the ego – and thereby simultaneously suppressing their dangerous mass instincts. Even left-wing policies, suggests Curtis, are moulded to cater for people's individual desires, as capitalism has learnt to do with products.

But now we may be about to revert to the old, pre-Freudian, pre-Bernays ways. We may be about to enter a future where the concept of the self is fragile, due to the shifting impermanence and arbitrariness of the surrounding cyber-world, a future where the boundaries between our bodies and 'out there' are blurred. This future might also be one in which IT-based products can trigger or suppress emotions by manipulating the brain landscape as directly as drugs, but remotely and unbidden with far greater precision. If so, then *for the first time ever* the importance of self-fulfilment and status might fade. The sense of self will be obsolete, no longer fighting against the tide of relentless retraining and re-invention, no longer seeking a sense of purpose whilst unemployed, and no longer working to gain things that you no longer want. Instead you will surrender wholesale to the once-occasional passive reception of the senses, interacting with and to the ebb and flow of inputs and outputs from all other fellow-beings via your cyber-world. You – only there will no longer be any 'you' – are now simply a consumer of technology.

The technology of work in the future, if it exists at all, looks set to become increasingly bio-dominated, from the computers we use to the fuel that powers them. The degree to which we work will depend on how efficient that work is, and how sophisticated the technology. But more fundamentally still, by reflecting on what we will be doing in the future, we are exploring the values and agendas of forthcoming generations. All round, it seems likely that traditional boundaries in our lives will be breaking down – between home and work, work and leisure, work and retirement, one generation and the next, and even roles within a family unit. But up to now it has been those very boundaries and stages in our life-narrative that have defined us. The questions and possibilities that have arisen from thinking about the future of work and leisure boil down to the simple yet important question of the extent to which our successors will see themselves as individuals, how robust their sense of self will be, and thus the degree to which they may be manipulated.

If as part of the workforce we need to change all the time, to become reactive rather than consistent as a persona, then our clear-cut sense of who we are might start to unravel. One possibility is that as a substitute for the traditional status of work hobbies might become more important. But then we would need clear time for them to become like alternative jobs. Or we could develop a second-hand ego, by viewing fiction or indeed recorded facts. Then again, we could develop still further the kind of public ego, the collective identity rampant at football stadia, and lose the sense of self altogether by 'going real' en masse from time to time, or perhaps do the opposite – seek oblivion in the safer, more familiar cyber-world, or with drugs. In short we will need to cope with a void, the loss of a sense of self previously derived from work. In order to understand how the self of the future might be assembled, and thus how easy it might be to control, let's start at the very beginning: with genes.

5

Reproduction: How will we view life?

The most important thing in life is life – both sustaining it in ourselves for as long as possible and having children to attain immortality by proxy. Now advances in biotechnology are opening up new avenues. The 'posthuman' individual, a genetically modified human being, may well become reality before this century is out. Gregory Stock of UCLA, adviser on biotechnology to Bill Clinton, predicts:

A thousand years hence, these future humans – whoever or whatever they may be – will look back on our era as a challenging, difficult, traumatic moment. They will likely see it as a strange and primitive time when people lived only seventy or eighty years, died of awful diseases, and conceived their children outside a laboratory by a random, unpredictable meeting of sperm and egg. But they will also see our era as the fertile, extraordinary epoch that laid the foundation of their lives and their society. The cornerstone will almost certainly be the reworking of human biology and reproduction.

But such promise also casts a shadow. As Francis Fukuyama, Professor of Political Economy at Johns Hopkins University, warns: '. . . the most significant threat posed by contemporary biotechnology is the possibility that it will alter human nature and thereby move us into a "posthuman" state of history.'

For a long time now genes have been big news. Still, as bio-high-tech increasingly seeps into everyday life ethical dilemmas mushroom. As I am writing this, today's papers cover no less than three stories relating to genes. One is the story of Charlie Whitaker, a young sufferer of an anaemia that gives him a life expectancy of less than thirty years, who is being denied treatment that could save him: his illness could be cured by appropriate replacement cells that would then proliferate in his blood. These early-stage stem cells can come only from blood in the umbilical cord, and have to be as closely tissue-matched as possible. So Charlie's parents need to have another child – with the right genes. The technique (preimplantation genetic screening) is available to select an embryo resulting from IVF to ensure that Charlie's mother becomes pregnant with a brother or sister who will save his life. But the British Human Fertilisation and Embryology Authority are denying permission, fearing the prospect of giving the green light to 'designer' babies.

On turning the page of the newspaper, I read of the tragic case of a young woman recovering from an ovarian cancer that has left her infertile. However, prior to her chemotherapy treatment, she and her partner underwent IVF to ensure that they would still be able to have children. Now that the couple have split up, the still-fertile ex-boyfriend has withdrawn his permission for the embryos to be implanted; indeed, according to current legislation, they may have to be destroyed.

Finally, I see that scientists have discovered a gene linked to aggression; this could be important news for one Stephen Mobley, whose defence against his imminent execution in the USA is that he comes from a long line of rapists and criminals, and hence his genetic make-up leaves him helpless. So, as genetic profiling becomes more sophisticated and more common will we all soon be able to blame our genes for any transgression? And if not, where and how will we set the defining criteria for drawing the line? It is hard to imagine a clear boundary, on one side of which there will be those who can turn over accountability to their biological inheritance, whilst on the other there will be the rest of us who, despite our genetic destiny, have to face the music.

As we look into the future, the manipulation of DNA will increasingly have implications for how we reproduce; but the new technologies will also touch our own little span of mortality, regarding our expectations of life, and have a dramatic influence on how we deal with disease and ageing. To most people, including molecular biologists themselves, the rate of progress has been startling. So fast have things moved that it seems we are already living in a sci-fi world where the previously most non-negotiable of tenets are no longer inviolate. Just how far along are we at this very moment?

At the centre of every cell of your body, except egg or sperm cells, lurk twenty-three pairs of large pieces (chromosomes) of deoxyribonucleic acid (DNA). A gene is any one particular segment of DNA that will eventually trigger the production of a particular protein. Amazingly, however, only a puny 3 per cent of DNA carries out this central job, and although some portions will be regulating other genes, the vast majority is 'junk'.

Another surprise is the remarkably small number of genes that we humans possess. The lowly worm C. *elegans* is made up of only 900 cells and has some 20,000 genes whilst we, whose brains alone are composed of 100 billion neurons and ten times as many 'glial' cells, have at the very most merely four times as many genes. When the human genome was completely mapped in 2001 we turned out to have only between 30,000 and 80,000 genes. And yet even though we have this relatively small number of genes, most of them shared with other animals and even plants, there are already as many as 5,000 known genetic disorders. Traits – anything from breast cancer to depression – are not contained completely and independently within a single gene: for example, there are already some ten or so genes 'for' Alzheimer's disease. And even in the rarer cases where only a single aberrant gene is the cause of the problem it is hard to imagine that such a condition is actually trapped within the strands of DNA. Even Gregor Mendel's purported gene 'for' yellowness in peas was actually the expression of an additional enzyme to destroy the green pigment chlorophyll.

That said, surely the more we know about which range of genes is linked, however indirectly, to which illness the better. After all, such information can but help with improved diagnosis and eventually with the design of new types of treatment. One way to discover new genes that are related to disease is by studying the genetics of populations. It is critical to be able to scrutinize large groups of humanity who have existed as relatively isolated communities, such as the citizens of Iceland, for whom the caprices of history and geography have ordained far less mixing of the overall gene pool than usual. On this smaller genetic stage there is far more chance that any variation in a particular gene can be more easily matched up with any particular disease. A second approach is to identify variations in genes due to mutations over centuries of evolution. These variations are known as 'polymorphisms'; since they are usually due to a change in only one rung (nucleotide) of the ladder-like structure of DNA the cognoscenti refer to them as single nucleotide polymorphisms (SNPs, pronounced 'snips').

Already this type of comparison between genetic deviation and illness has proved useful: gene mutations have given valuable insights into developing screens for individuals at particular risk of a range of conditions from breast cancer to heart disease. For example, a change (mutation) in one particular gene (p53) is linked to over fifty forms of cancer – in fact a colossal 90 per cent of all cervical cancers, 80 per cent of all colon cancers and 50 per cent of all brain cancers are associated with an aberration in this single gene. Indeed, so important is this gene in relation to cancer that in 1994 the journal *Science* voted it 'Molecule of the Year'.

Easy to see, then, how genetic tests will become increasingly part of our lives. They will soon, routinely, be detecting whether a host of diseases are already present in your body, or likely to occur, whether you might be a carrier, whether you are at increased risk of a complex condition such as heart disease or cancer, and what specific reactions you might have to medicines, food or environmental factors. Along with establishing your paternity and genealogical record, your individual past, present and future will soon be expressed in cocktails of genes which interact with your environment to give, well, you.

Needless to say, such unprecedented power of diagnosis and awareness of the risks will radically change how future generations think and live. Eventually it will be normal for you, or any third party, to refer to a biochip read-out of your entire genetic profile, thereby allowing personal screening for side effects and possible differential responses to drugs. The downside, of course, is that there is a vast new potential for individual discrimination and inevitable issues over insurance premiums. Moreover, as different types of tests become rapid, easy and inexpensive so debate will increase over whether they should be sold directly to the public. Whereas we are comfortable with home pregnancy tests, imagine a DIY test for Alzheimer's disease . . . And if pretty much everyone has to get used to living with risks that, with the benefit of early warning, they might be able to reduce perhaps, but not abolish completely, then that could make for a more edgy and self-conscious society. Some might argue that we know more now about the risk of motor vehicle accidents than forty years ago but this has led to a safer society with more choices. Then again, the risk we take when driving in a car is a short-term, acute one, and we take appropriate, direct measures for improving the safety of each journey, which is not quite comparable to the far more indirect and long-term link between risk factors and a disease.

This living with risk will be even more problematic if you discover that you have an aberrant gene 'for' some brain disorder or mental function, or if you wish to exploit normal gene function to boost the cerebral prowess of your progeny. The special difficulty here is that, when it comes to the brain, to those most elusive and fascinating mental traits, genetic provenance is particularly tenuous. Take, for example, the gene that would surely be the most coveted of all by prospective parents, that for IQ. Even if a single gene for IQ were to exist, it seems its contribution would not be a major one. Roger Gosden, in his book *Designer Babies*, has actually put some numbers on the concept. On a scale from 0, as a purely environmental causation, to 1, completely hereditary, IQ scores show about a 0.3 score of heritability. A very major single gene, were it to be discovered in relation to IQ, would probably account for at most about 5 per cent of the trait so its presence would contribute to some 1.5 per cent of the final IQ scale – if you had an IQ of 100, this gene would then boost it to 101.5. Gosden advises parents of the future to invest any money they might have in their children's education rather than in dreams of genetic enhancement. On the other hand, most people would put the value of heritability for IQ much higher, probably about 0.6. In any event, education seems to add about five IQ points, which is slightly more than a single gene with a heritability of 0.6 would. However, such calculations have only limited value; the real issue is that the gene could influence how you interact with your environment, and the old line drawn between 'nature' and 'nurture' is becoming so blurred that it really is not very helpful.

In 1994 Richard Hernstein and Charles Murray reached conclusions that were unpalatable to virtually everyone: they showed that much of the variance in intelligence was due to genes, and therefore had a racial bias. But the study should now be revisited. In particular, the finding that African-Americans had a significantly lower IQ than whites seems explicable, not, after all, by their genes but by environmental factors. The psychologist James Flynn has subsequently shown that IQ scores have been rising over the past generation in nearly all developed countries; the IQ gap between white American children and the children of non-Caucasians, and indeed all immigrant groups, could well be closing, as improved nutrition and education leave their mark, literally, on the brain.

In any case, there is the enduring question of what we mean by 'intelligence', and how such a vague concept can ever be realized by the mindless mechanics of physical neurons in your brain. One example of the yawning gap between behaviour and thoughts, on the one hand, and genes and proteins, on the other, cited by Francis Fukuyama, comes from the neuroscientist Joe Tsien. Tsien found that, by eliminating or adding a certain gene, he could respectively worsen or improve memory in mice. Could he have discovered the gene that codes for the protein 'for' memory?

Hardly. This gene is one of several controlling the availability of a protein that happens to be involved in a particular chemical communication system in the brain, one which allows large amounts of calcium into the neurons. But the big problem is that calcium is like grapeshot inside a cell: once it has gained entry it will find many different molecular targets, and result, directly or indirectly, in many different effects, some beneficial, some toxic. Memory will not be the only mental property to be affected. And whilst these changes may well be *necessary* for memory to take place, they are far from being *sufficient*. Many other factors will also determine whether, and how well, the brain retains a memory. Remember that the protein made by a particular gene does not have a one-to-one correspondence with a mental trait. In most cases, that protein (and the gene that expressed it) will be involved with many traits, and, conversely, any one trait will involve many proteins (and genes).

A similar situation arises with the highly controversial 'gay gene'. At first glance there does indeed appear to be a strong hereditary component in sexual orientation. (For example, one study has shown that 52 per cent of identical twins with a gay twin are themselves gay; but for non-identical twins, who, of course, share only half their genes, that number falls to 22 per cent.) But how might we then explain the fact that people often switch their sexual preferences as they go through life? Even if there were a gay gene, there is no saying whether or not it is being switched on or off at different times of life by different environmental factors. Moreover, the influence of 'environment', since this includes the macro, complex events of the outside world, is impossible to deconstruct and trace to immediate molecular phenomena outside the walls of brain cells. Indeed, being gay could well have much in common with heart disease. There is a very small number of people with familial hypercholesterolaemia who are genetically determined to have heart disease, and do develop it, but there are far more sufferers whose heart disease is the result of an interaction between genes and environment. Meanwhile, a small number of other individuals have genes that predict a very low risk of heart disease but drink or smoke heavily, are obese, or do not exercise – and have heart disease. The relationship between genes and sexual preference could be similar: for most people it is a combination of genetic predisposition plus environment, including a component of choice.

But sexuality, and indeed gender stereotypes, do seem to have a strong 'natural' basis. I remember having a drink with friends whilst their son, oblivious to us boring adults, played with his trains; his sister, however, also of pre-school age, force-fed me peanuts, already engaged in a three-year-old's version of nurturing. And a more formal, unambiguous demonstration of gender-linked tendencies can be seen on a video of pairs of 7- and 8-year-olds ostensibly passing time in a waiting-room that contained a range of toys. The little boys start playing immediately, competing with each other in various games, and restricting conversation mainly to directives and scores. For most of the pairs of little girls, however, the toys are secondary to establishing a relationship, finding out how many siblings they each have, what they are going to be when they grow up, and so on. And almost every parent with whom I have discussed this issue will avow that their sons have a different way of behaving, somehow 'wired-in', that is distinct from that of their daughters.

More formal studies show clear differences in attention span, verbal learning, construction ability and many other skills between girls and boys. Arguably, the book *Men are from Mars, Women are from Venus* was such a bestseller because much of the adult gender-based behaviour it described was all too familiar. If sexual orientation is just part of the general gender-related features of mental function, then it is clear that some biological switch must be thrown, if not in a single gene then very early on in life, to determine it.

The current consensus is that brains are sexualized within the womb according to the levels of exposure to the male hormone testosterone, linked to the Y chromosome. This testosterone will have many effects both on the structure of the brain and its subsequent operations. A little like the protein that, in a very indirect way, affected memory, so varying levels of testosterone are necessary but not sufficient for determining an individual's sexual orientation. Many other factors, including other genes, are involved, whilst testosterone itself contributes to the expression of other traits, such as aggression.

So what sense can we make, after all, of the notion of a gay gene? In industrialized society only 1 per cent of men are exclusively homosexual, whilst some 5 per cent are bisexual. Yet bisexual men have double the ejaculation rate of heterosexuals and a larger number of partners, including more women. In his lifetime a bisexual man is therefore likely to inseminate more women than a heterosexual man. At the moment bisexual men and women are in a minority because a variety of factors, including the risk of disease and physical or verbal attack from homophobes, militate against activities other than heterosexual ones. In the future, however, if the constraints of both prejudice and disease are removed, the numbers of bisexuals will escalate, according to biologist Robin Baker, due to an unfettered dissemination of their gene profile via their more active and diverse sex lives.

Nonetheless, one prediction is that in the future couples could vet the genes of their embryo for particular traits; homosexuality may prove to be a characteristic to screen against on a par, say, with baldness – harmless enough but not particularly desirable. On the other hand, it might equally well be the case that bisexuality becomes the norm; since the nuclear family may be due to fragment completely, and if AIDS and other related health hazards can be removed, an ambiguity of sexual orientation, a blurring of conventional roles, could well reflect the increased and more generalized tendency for a less rigid, less stereotyped sense of self.

One final example of an alleged single gene 'for' a complex mental trait – indeed one that featured in my newspaper today – is of the gene 'for' crime. The idea that criminal tendency could be inherited gains credence from a study of 3,586 twins from the Danish Twin Register: identical twins had a 50 per cent chance of sharing criminal behaviour compared to a risk of only 21 per cent among non-identical twins. Similarly, a much studied Dutch family, with a repeated history of violence, would suggest that the problem might be genetic.

One approach to the issue of the gene for violence lies in examining the relation of genes to brain chemistry. The tendency towards violence could be due to an impairment in certain genes controlling a particular enzyme that, in turn, regulates the availability of a particular chemical messenger, a transmitter. This hypothesis is all the more attractive in view of the fact that mice with a similar chemical defect also seem to be very violent. However, this situation is just like that of the protein related to memory in neuroscientist Joe Tsien's experiment, or like the contribution of testosterone to sexuality. Once again there is a component chemical that is all-important in the final function or dysfunction, just as a sparking plug is vital to the function of a car. But there is far more to a car than a sparking plug, and indeed a precarious sequence of events between ignition and the movement of the vehicle. So it is with genes and brain function.

Although genes are identifiable as unambiguous entities, they are not at work incessantly and autonomously: a variety of constantly changing influences will switch them on or off from one moment to the next and will thereby determine whether a particular protein is to be made, and indeed exactly what type of protein it actually is. Many factors, such as the age of cells, their development and, most importantly, their immediate chemical environment, will determine whether a gene is activated at any one time. The seething micro-environment of chemicals within the cell will itself be influenced by macro-environmental factors, chemicals that seep into the fluid-filled spaces between the brain, which are in turn influenced by what is happening in the rest of the body, and beyond, in the outside world.

In the case of the brain, a vital element will be the ceaseless communication from neighbouring cells, as transmitter signals stream onto each neuron. Since we have some 10^{15} brain connections, the physical junctions between one neuron and another will outnumber the entire number of genes by some 10^{10}. And the type and amount of transmitters in active service at any one moment will be determined by the operations at that particular instant within large-scale circuits of neurons that make up different brain structures – which in turn will be influenced by what is happening both within the rest of the body and ultimately to the individual as they interact with the outside world. The relation between brain functioning and the chemicals that underpin it, and even more so the genes that make the proteins that make those chemicals, is as convoluted, remote and capricious as in the old verse 'For want of a nail', which traces how a nail missing from a horseshoe led to the eventual loss of a kingdom. Yes, there was a causal relationship, but that link between the loss of a nail and the result of an armed combat is so indirect and tenuous that it is surprising that it was identified at all.

This subtle dialogue between nurture and nature can be seen in action in a recent experiment by Colin Blakemore and his colleagues in Oxford: they were investigating the influence of the environment on the disease Huntington's Chorea. Huntington's Chorea is a severe disorder of movement, characterized by wild, involuntary flinging of the limbs; hence its name, from the Greek for 'dance'. It is an example of a disorder of the brain that, unlike Alzheimer's or schizophrenia or depression, can be blamed on a single gene. Normally, within this gene, one particular sequence of three (triplet) component chemicals repeats itself over and over again, up to twenty times. But if the number of such repeats increases to thirty-nine, then the disease appears in an individual in their mid sixties; forty-two repeats and the patient will be afflicted by their forties; fifty repeats and they will be ill by the time they reach their thirties. The greater the number of repeats the earlier in life the disease strikes. Surely here, then, the cause of the disease is entirely genetic and we can justifiably talk of the 'gene for' Huntington's Chorea; could this disease be the exception that tests the rule of a constant interaction between nature and nurture?

Blakemore and his colleagues were working with 'transgenic' mice, so called because they had been engineered to have the faulty gene 'for' Huntington's Chorea: their performance in tests of their limb control would therefore be certain to grow worse with age. But surprisingly, despite the clear relation between gene and illness, it turned out that the mice could be protected enormously by environmental stimulation, by exposure in early infancy to mouse toys – playthings which provided an extra opportunity to interact and explore. The performance of the mice was nowhere near as dramatically impaired, nor did the disability start as early, if they had experienced a certain kind of lifestyle.

We can see from this type of experiment that a gene is not an independent or autonomous agent. Instead, it is a component of the brain that may or may not be triggering production of a protein at any one time. As such it works within the context of the whole brain. When a gene is activated it makes a related chemical (messenger RNA), which acts as a template for the construction of the specific sequence of amino acids that will eventually constitute a much larger molecule, a protein. But segments of this mRNA can be spliced out before the particular protein is made, so that some 'edits' give different products from others. The number of different combinations of mRNA segments multiplies enormously the potential for what each gene might 'do'. In one case, in the fruit fly, a single gene could be the 'gene for' a staggering 38,000 different proteins! And then the same protein can be transformed into something different and varied after manufacture: depending on whether it was incorporated as a component into some aspect of the more complex machinery of the cell, or was perhaps coated with sugars to form up to any of some 200 chemical groups, its 'function' would change. Small wonder that despite the nugatory number of genes we have – remember, 80,000 at most – our bodies are making and using millions of different proteins.

This discrepancy between number of genes and number of proteins, not to mention all the nested anatomical and functional hierarchy of the brain in which we have just glimpsed those proteins at work, accounts for why it is usually naive and misleading to expect to extrapolate from a single gene to a complex mental trait, such as IQ, homosexuality or criminality. Even when the 'cause' is inherited, many genes are at work and in incessant dialogue with non-genetic factors in their environment. Hence we should not be too surprised to learn that different genes work differently in different species – for example, a mutated, breast-cancer gene causes mice to die as embryos, whilst in humans everything is normal until well into adulthood. Even in yeast, which has only 6,297 genes, the pattern of protein manufacture of more than 300 of those genes will alter as a result of a single mutation in a single gene. Contrary to some current thinking, then, genes do not set any agenda or have minds of their own, nor do they perform one job only in autonomous isolation.

Nonetheless, just because genes have to be put in their place their huge potential for new types of medication is by no means negated. By understanding how any one gene triggers the manufacture of a particular protein, new types of drugs can be developed that are proteins themselves. As well as insulin, which has been in use for many years for early onset diabetes, there are now new treatments using human proteins for wound healing and blood-vessel growth. Currently there are only 400 to 500 molecular targets for drugs; in the case of mental dysfunctions, the main treatment strategies work at any one of the different successive stages (synthesis, action and removal) to change the final availability of the chemical messengers, the transmitters. However, during the next decade, because of the potential advances in the genetic sciences, there will be a massive explosion of new types of targets, to some 4,000. These new drug targets will enable diagnostics, protein replacement and the use of drugs (monoclonal antibodies) that combat the undesirable effects of certain proteins by direct molecular interaction with them.

But on the other hand, we won't be able to trace easily how a normal function is 'caused' by a particular gene, just because we can correct a faulty component and therefore help cure a dysfunction; just as you will never work out how a complete engine functions by contemplating a sparking plug placed by itself on a table, simply because the replacement of a faulty sparking plug will enable you to drive your car again. We can work backwards from a dysfunction to the cause, and fix it. But we cannot work forward from a single factor or component and expect to understand the whole system. There is therefore a massive difference between gene-related treatments for certain illnesses and vague hopes for enhancement of a complex mental trait. This point is a really important one, if we are to evaluate and predict how far the new advances in genetics are going to take us in our lives in the 21st century. Some innovations are just around the corner whilst others are pure fantasy, for good or evil.

Imminent gene-related development could change our lives in the very basic area of paternity and fertility testing. Paternity testing is already in extensive use for resolving situations where the identity of the father is a critical issue. The basic principle is to reveal the particular, unique pattern of the DNA that flanks our genes. This technique is 100 per cent accurate, and because it is so reliable the likelihood – an urban myth places it currently at 10 per cent – that a man may be unwittingly 'fathering' another man's child is zero.

But why should all fathers not be entitled to the same assurance? By making the paternity test universal and obligatory at the time of birth society will be able to ensure that every man knows exactly his responsibilities, or lack of them. In his visionary yet chilling *The Future of Sex*, Robin Baker reasons further that it would be fairest to tax fathers, irrespective of whether they stayed with their families or not. After all, a child costs the same sum to bring up whether the father is not there at all, a commuter dad who sees his child for a few hours a week, or there all the time. By introducing a universal child tax for biological fathers, Baker believes, we would have a much more equitable state of affairs. All genetic children would have an equal financial share from their father, according to their age but irrespective of whether or not they were his first or second family. Only in this way, Baker asserts, will we have a just society where no mother is simply abandoned nor any man duped; and it is only possible due to the science of genetic fingerprinting.

A possible downside of this arrangement is that a woman could ensnare a wealthy man or two as a meal ticket: if paternity testing was compulsory, proven paternity would guarantee an income for eighteen years or so. A problem for a woman relying on this scheme would be the 'time wasted' seducing such well-off candidates for fatherhood without knowing definitely whether she was in the fertile part of her cycle. Hence a further reason – up until now – for the nuclear family: the man and woman need to stay together so that they can copulate on a monogamous and genetically accountable basis over an extended time frame, since neither can be exactly sure when the woman is fertile. Because sperm can live in the female's body for up to five days, we need to anticipate ovulation up to five days in advance, and as yet that has not proved possible. But if, as predicted, such testing was to become infallible, then for the first time ever an unwanted conception would be more costly for the man than the woman. In order to justify the current nuclear family, in reproductive terms, the woman must gain help from the man in raising that family, whilst the man must be unable to tell if the woman is fertile. In Baker's scenario, thirty or so years from now obligatory paternity testing and an infallible fertility predictor would, in combination, remove both those factors.

But the genome era will affect far more than our reproductive habits. One of the most immediate benefits could be new treatments for disease based on stem-cell therapy. Stem cells are the basic, generic cells from which other, more specialized cells develop, according to the biological environment in which they are placed, be it bone marrow, or heart muscle or brain. These highly versatile and premature cells can be generated artificially at later stages in life by cloning: the process whereby the nucleus of an unfertilized egg (i.e., most of its own DNA) is removed, and the DNA from an adult cell is introduced instead. This cell and the empty egg are then fused together such that they divide to yield 'stem cells', from which eventually nerve cells, heart-muscle cells, skin cells, bone and blood cells will all be elaborated. Clearly replacing organs with these cloned, forerunner cells is a very attractive idea: it circumvents the ethical difficulty and time delay of obtaining donor organs (the donation itself is uncertain, the timing completely unpredictable, and the relatives of the deceased may have to give permission when already in considerable distress).

Of the many applications of stem-cell therapy, a recent project that particularly captured journalists' attention was a technique pioneered by the biotech company ReNeuron, who plan to implant stem cells into areas of the brain damaged by stroke or neurodegenerative disease. Data from animal experiments is promising, and the technique offers hope for disorders for which there is currently no effective treatment. However, it is hard to estimate the damage that may be caused to brain tissue by the implant as it is pushed into its correct location deep within the brain, and it would be equally difficult to measure how well these alien cells could eventually replicate the intricate wiring of the normal cells in an adult system. No one will volunteer to act as a control – to undergo a sham brain operation, whereby the conditions of anaesthesia, and indeed the whole surgical operation, are identical to treatment, but an inert substance is implanted instead of stem cells. Even when the cells are implanted, and in the correct place, it will be hard to know how to regulate what they then proceed to do.

Take, for example, Parkinson's disease, a neurodegenerative condition which causes a sufferer difficulties in moving, as well as severe muscle rigidity and tremor. In Parkinson's disease a specific group of cells towards the very base of the brain start to die, and the amount of the particular chemical (dopamine) that they use to communicate globally with remote systems and structures all over the brain starts to dwindle. Although stem cells could, in theory, replace the dying neurons and restore the levels of brain dopamine to normal, there is still a difficulty. Again, a transmitter, just like a gene, can contribute to more than one final function or trait. Although deficient levels of dopamine underlie the lack of movement that characterizes Parkinson's disease, excessive amounts of this same chemical can affect another brain system altogether and end up causing schizophrenia-like psychosis. If stem cells are implanted in the brain and start to release dopamine, how will we regulate the amount of transmitter that they release, so that the patient does not end up perhaps with better movement but now with crippling and terrifying hallucinations?

Another problem is that it is the business of stem cells to divide and proliferate into still more cells – so how might we ensure that the process does not get out of hand and result in a tumour, which is, after all, an inappropriate division of cells? There could be ingenious ways around this difficulty, such as ensuring by genetic engineering that the stem cells in question will divide only at temperatures a few degrees hotter than will ever occur in the real brain; but clearly the procedure needs to be improved. Then there are the non-trivial considerations – especially among the elderly – of the risk, expense and sheer unpleasantness of brain surgery. In summary, stem-cell implants into the brain may offer an exciting new development in the potential treatment of brain disease, but are far from ideal.

Another new approach to disease, based on the advances of

molecular biology, is gene therapy. Gene therapy leapfrogs the unwanted product of a defective gene by inserting a new, well-behaved gene that can produce the 'right' protein into the chromosome. As more and more genes were identified over the last decade that could be linked to various diseases, so this approach seemed as exciting as it was obvious. However, things have proved far more awkward than initially hoped: one big difficulty lies in simply gaining access to the faulty gene, locked inside the nucleus at the centre of almost every cell in the body. And it is even more problematic to get the normal gene into the cell in a functional form. For many genetic disorders the defective gene is totally non-functional and makes no protein, but even in these cases gene therapy has not worked, due to the problems of entry and regulation. One option is to extract bone marrow, treat it with the engineered stem cells, and replace it in the whole person: over time the cells in the bone marrow will reproduce themselves, and gradually spread through the body. However, most effective would be a treatment whereby the new DNA could penetrate cells that did not have to be exported from and re-imported into the patient's body. As an alternative to using viruses, which are rebutted by our natural defence mechanisms, a new approach called biolistics is being developed: DNA is mixed with small metal particles, such as tungsten, then fired into a cell at high speed.

Other alternative approaches for gene therapy involve DNA encapsulated in liposomes (little parcels of fat), or injected into liver or muscle cells bound to calcium phosphate, which will help some cells take up the new DNA and express appropriately engineered genes. But it is proving very hard to get gene therapy to work, and it seems to be a less than ideal route for developing new therapies of the future. Rather than work with cells that are already set on their course to function in certain ways in the body, we could engineer the DNA in the infinitely fewer cells from which a person has yet to originate – in one egg or sperm cell.

Already sperm can be screened for gender selection. Male (Y) sperm are lighter, so they can swim better through thick solutions of albumin. Hence some claim that 'male-enriched' sperm can be made by filtering the ejaculate through such a solution. A female (X) sperm has more DNA (2.9 per cent), so it takes up more dye and thus gives off more fluorescence if irradiated with a laser. Here then is another means to detect gender prior to conception. Yet a further technique is to add molecular blockers (antibodies) specific to one or other gender that will then inactivate the sperm so targeted: X antibodies in the sperm would increase the chance of conceiving a boy, Y antibodies the chance of a girl.

In any event, conception outside the womb has opened up a wide range of possibilities that do not have implications for future generations but could still radically change the way we live in the future. Since 1978, when Louise Brown was hailed as the first 'test-tube baby', some 68,000 babies have been born in the UK alone using 'in vitro fertilization' (IVF), the fertilization of an egg outside the womb. Once the process has proved viable, and the fertilized egg (a zygote) has grown to some eight cells (the embryo), it is implanted in the womb.

The pioneering fertility expert Robert Winston used a gender screen to ensure, in one particular case, that a future embryo in an IVF procedure was a girl: as such she would carry an XX, and thus avoid a rare disease carried in her family and linked to the male (Y chromosome) line. Hundreds of diseases are linked in this way to the sex genes, and other genes relating to diseases, such as cystic fibrosis, are found on other chromosomes. So preimplantation genetic diagnosis (PGD), which Winston started in 1989, now screens for other diseases too. One or two cells are sampled a few days after conception, at the eight-cell stage of development of the foetus: the procedure has the potential to reveal any of the thousands of 'genetic' diseases that have been recorded so far, and therefore may well become more and more frequent, perhaps eventually even routine.

Once a screen shows up a problem, then currently that embryo is not used. Alternatively, as in the case of the anaemia victim Charlie Whitaker, screening could be used to select positively one embryo over another. And as we progress from screening for diseases such as cystic fibrosis, or Down's syndrome or spina bifida to screening for hardness of hearing or poor bone growth, where do we draw the line? Surely, as with all reproductive issues, this comes down ultimately to an informed choice made by the parents, and ultimately by the mother, as opposed to by doctors or governments or churches. Yet the big question still remains: whether the opportunities afforded by such powerful technologies will instil a new pro-eugenics mentality.

In the future techniques could also prove popular not just with infertile couples but with women who wish to establish their careers before having a child and yet still have a baby at the biologically optimum time. Below the age of twenty-five 95 per cent of women will conceive within six months of unprotected sex, whilst above the age thirty-five years of age less than 20 per cent do. We now know that the age – not so much of the mother as of her eggs – is critical for the chances of a healthy baby, and indeed of a pregnancy at all. Imagine therefore a scheme whereby a woman had eggs frozen at her biological optimum, say when she was eighteen years of age: she could then postpone pregnancy for as long as she wished, and still give birth eventually to the healthiest baby possible.

Now imagine that the young woman in question has her Fallopian tubes blocked so that she is infertile. Men could also take advantage of the scheme and have their sperm, along with the stem cells that make it, frozen, followed by a vasectomy. In so doing, men would be protecting themselves from the machinations of devious women aiming to trick the man into fathering a child. At last, half a century or more after the contraceptive pill was first developed, the link between sex and reproduction will have decreased to zero.

The implications of this are profound. If sex is *just* for pleasure, with no risk whatsoever of pregnancy, not even remote and covert at the back of our minds, then perhaps we shall engage in it with even less commitment than currently. A good comparison might be with the gay community: homosexuals have close emotional bonds and friendships but far more sexual partners. Factor in the cyber-technology we were looking at earlier and it could well be that the act of sex becomes, like much else in the future, a passive hedonistic experience, where you 'let yourself go'; meanwhile, personal relation-ships, where you have a clear identity and a clear role, be it with lovers, ex-lovers, friends or family, no longer include the escapist times of utter abandonment. Your personal life, with its plots and subplots, its complex inter-relationships, track records and histories, becomes a distinct and separate component to your existence – or, as a result of the new passive lifestyle of hyper-stimulation, starts to fade away . . .

In this increasingly compartmentalized existence, reproduction will rely more and more heavily on 'in vitro' techniques, like IVF, which take place outside the body. Further refinements of IVF can already, and will even more so in the future, circumvent causes of male infertility other than blocked ducts. If the problem is that the sperm cannot swim or that they are too few in number, then they can be injected straight into the egg, a technique known as ICSI (Intra-Cytoplasmic Sperm Insemination). And if sperm are not being manufactured at all, there is ROSNI (Round Spermatid Nuclear Injection), where precursors to sperm (spermatids) can be introduced into the egg and still fertilize it, using techniques now available for screening prior to implantation of embryos in the womb. The biggest reservation here is that the process avoids the normal test for sperm fitness: vigorous swimming towards the egg to establish the 'winner', in Darwinian terms of survival of the fittest. All these technologies for intervention at or around conception will undoubtedly enable us to live and work in a different way, and to have 'healthier' children. But where does health end and a quest for physical and mental perfection begin?

'The elements so mixed in him, that Nature might stand up and say to all the world, "This was a man!"' says Mark Antony of Brutus in Shakespeare's *Julius Caesar*. Surely it is our very imperfections, and the ways we deal with them both in ourselves and others, that have kept poets, novelists and historians in business for millennia. If screening is taken to the limit, then a homogeneous society where everyone, but everyone, is super-healthy and mentally stable (barring any traumatic life events) could occur; a situation that is hard to imagine from our current perspective of a humanity with a mass of imperfections. The extreme, especially if coupled with the less taxing cyber-relationships we have been discussing, will inevitably be humans with 'less' Brutus-type human nature. On the other hand, we have already seen that genetic technologies may well transform physical health below the eyebrows; but when it comes to the brain the elimination of diseases, let alone the 'enhancement' of normal mental prowess, will be far more difficult.

Yet it is not just our future progeny for whom genetic intervention will affect health, life expectation and relationships. Those of us who will soon be the new grey generation may not remain untouched by the hand of molecular biology. We saw earlier how a healthier lifestyle and dramatic modifications to it, such as calorie restriction along with far more effective monitoring, would stave off many of the diseases with which we are currently familiar. However, the more developed the society in which we live the more hereditary factors appear to dominate in determining our lifespan, over nutritional and other environmental issues.

One idea is that ageing is due to dysfunctional senescent cells releasing potentially toxic substances on to other cells as they slowly deteriorate. One anti-ageing strategy, then, would be to introduce a killer gene that became active within a cell once that cell was showing signs of sickness: this sudden death would be safer for the body than a slow demise that could affect all the neighbouring cells. Another possibility is to exploit the age-defying actions of the enzyme telomerase. As we saw in Chapter 2, this enzyme can prevent the ends of chromosomes in the nucleus of each cell from fraying, by preserving the shoelace-type caps, telomeres, that otherwise deteriorate with age. Normally telomerase operates only in stem cells, cancer cells, and sperm and eggs, where – for good or ill – the chromosomes need to be in tip-top condition; the idea would be to engineer all the ordinary (somatic) cells of the body so that their genetic profile could remain equally pristine.

Although ageing might be tackled using genetic technologies, it does not follow that we would *necessarily* live for longer than the 100 years or so that seems to have remained a constant maximum throughout history – it's just that more of us would do so. But were a specific ageing gene to be identified then presumably modification to said gene might enable us to live, for the first time, for appreciably longer than ever before. It has been quite a few years now since Seymour Benzer identified the 'Methuselah gene' for fruit flies, which enabled them to live a third longer than their non-genetically enriched counterparts. However, it would be quite another matter for such a gene to exert its dominance in the complex bodies of humans. More likely, ageing is comparable to intelligence or sexual orientation: there is a complex genetic component, but one that is highly and continuously interactive with the environment. Ageing, like mental prowess, is not a simple single phenomenon but rather an umbrella term covering myriad events.

Nonetheless, a society in which most people lived to be 100 years old would be very different from the one in which we are living now. The first issue would be whether the older generation were active, or helpless and in need of constant care – and the proportions in which these two very different constituencies co-existed. Then there would be the predicament of a physically debilitated individual who still had an active mind, or the reverse: imagine being physically fit, but with a brain that could no longer register what was happening around you. In any event, career structure, political organizations, allocation of resources, family structure, retirement schemes, housing, work and leisure activities would all be skewed by a shift in society where the post-reproductive sector was the majority. But perhaps even that assumption of an end to fertility should be challenged. After all, we can now produce clones from adult mammals.

Within the next few decades the technique of cloning humans should be established sufficiently to satisfy the qualms of those who argue that as yet the results are too variable for complete confidence that the procedures can be considered safe and reliable. After all, the prototype clone, the sheep Dolly, was the only success out of 277 fusions of eggs with the genetic material, the DNA, from an adult udder cell. The reason why Dolly was such a breakthrough was because, until she proved otherwise, adult DNA was considered to be no longer able to re-enter into cell division. This discovery of how to switch adult DNA back on and clone from adult cells (far harder than from immature ones) opened the door to cloning dairy animals with high milk or wool yields, to which end most work on cloning has actually been directed. As well as the production advantages of a uniform, healthy stock, animal engineering can now make vital medicines and products for human health, such as the protein lacking in one form of the disease emphysema, or lactoferrin, the source of iron in mothers' milk.

Cloning really just means copying, and copying genes as DNA raises no ethical issues. It is only cloning people that raises problems. Accordingly, there are several types of cloning. 'Molecular cloning', of just a few genes, is mainly a research tool that need not bother us here. Nor will we spend much time on the cloning of embryos that will remain, for the most part, as embryos; stem cells from such embryos could combat disease, whilst embryos could also be cloned for IVF procedures so that a woman does not need to undergo the discomfort of repeated cycles of egg harvesting. Whilst the issue of using human material in this way as a source of spare body parts is far from trivial, most controversy is generated by the idea of cloning as an infallible treatment for infertility – be it cloning an existing child, a third party or indeed, as 7 per cent of respondents in a recent *Time* magazine survey would wish, oneself.

The immediate reflex objection to cloning, as to artificial insemination in the 1930s and IVF in the 1970s, is that it is 'not natural'. The standard rejoinder to this argument, often lodged against many scientific developments, is to question what is 'natural'. Taking an aspirin for a headache, for example, would not qualify, nor having a broken leg put in plaster, and certainly not having a heart transplant or an artificial heart. On the other hand, there are no medical reasons for intergenerational cloning; it is solely an issue of social preference, personal priorities and the choice of a small number of individuals.

A more specific objection, however, is that sexual reproduction in the normal way allows the possibility that the offspring will have useful traits for dealing with a new environment not apparent in either parent. Maybe, but no one is suggesting the whole of humanity switches to reproduction by cloning. It is hard to see how the continuation of our species on this planet will screech to a halt if a cloned baby is born to a desperate childless couple, or to a committed and loving same-sex couple. There are, in any event, already 48 million clones alive and well – identical twins; as geneticist Gregory Stock points out, the world does not seem the poorer for their being among us. The same twins could also stand as a counter-example to those objecting to cloning on the grounds that it is important to have genetic uniqueness. Moreover, there are many children who favour one parent very strongly, in looks, say, rather than the other; a cloned child, on the face of it, literally, would be no different.

These objections, and indeed the flights of vain fantasy, that arise in the cloning debate are based on the mistaken time-worn assumption – yet again – that genes are autonomous components of brain function, and that they will dictate exactly how a clone feels, thinks and acts up to the point of being a perfect simulacrum of the original. We saw earlier that genes have a part to play in brain operations but are far from being autonomous or predominant in the emergent mental traits. And at best, you could only ever have a cloned daughter or son – always you, the adult, will be separated from your clone by a generation, and that generation difference will mean a world of difference in culture, fashions, diet, health and education. You will hardly be more exactly like your cloned offspring than you would be like a child conceived conventionally who strongly resembles you, and certainly you will probably be more different in character from your clone than from your identical twin, born in the same generation.

So, in the end, what does the future hold? Once cloning techniques are perfected, and once adoption-type legislative precautions are in place, there will be no clear-cut grounds for preventing human cloning for infertile or same-sex couples, other than a possible diminution in uniqueness, as currently shown by studies of identical twins. Such people, desperate for progeny, would, I'm sure, have a very strong reply to those who simply dismiss the prospect because there is 'no reason' to clone humans and state that such couples 'can be helped in other ways'. It is getting easier, in the UK at least, for same-sex couples to adopt a child; even so, for many people, reproductive freedom is a personal liberty. And surely the critical issue, for adoption and cloning alike, is that the parents, heterosexual, homosexual or solitary celibate, prioritize the welfare and love of the child, take full financial responsibility for their actions, and do no harm to society at large or to the individual who is the child 'made' by cloning. A further application, resurrection cloning, whereby a dying relative is cloned, again, need not be intrinsically repugnant; once we grasp the fact that a clone really is *not* a simulacrum in all mental and physical traits of the original, then the horror, and along with it perhaps even the appeal, diminishes.

On a more realistic and practical level, leaving aside the need to develop cloning technology further before it could be used, there are other basic considerations. A surrogate mother must be found, if the commissioning mother is unable to carry a child, and more fundamental still is identification of a source of a ready supply of eggs to act as host to the alien, adult DNA which will be introduced to form the genome of the clone. Eggs must be available that are abundant and cheap. Payment of donors in the UK, as of surrogate mothers, is forbidden, but in the USA women can charge between $3,000 and $6,000 dollars for ten eggs. As it stands, this is not a technique that could easily become commonplace, say on the National Health Service, on financial grounds alone. For that very reason the problem may not be one of practical implementation but rather of creating yet further social divisiveness.

In any case, there is a good reason not to panic about cloning: as a solution for infertility it could become obsolete within this century. Some think that in the not too distant future it will be possible to take a cell from anywhere in the body, and deprive it of half its chromosome content – to become just like a sperm or an egg. Once gametes (sperm and eggs, and their equivalent in genetic terms) can be made in this way, from any cell, irrespective of the age or health or sex of the cell donor, then anyone would be able to have a baby with anyone else.

So, for example, a post-menopausal woman would still be able to have a baby, perhaps hiring someone as a surrogate womb if she herself was physically unfit. The father could be a matter of personal choice, using the usual range of criteria, or an anonymous donor from a sperm bank. However, Robin Baker goes so far as to predict a 'Gamete Marketing Board' whereby the consumer could select the DNA of football stars, or university professors or film stars. Irrespective of her sexual orientation, our heroine might even choose another woman, who could either supply the egg or, if she too was infertile, provide half the genetic material from a cell elsewhere in her body.

Similarly, male homosexual couples would be able to have a baby that is biologically all theirs. As for cloning, a donor egg would be needed, subsequently evacuated of all its original DNA. Half the genetic material would then be supplied by the haploid nucleus derived from any cell in the body of one of the couple, but depleted of half its DNA, whilst the other half would originate from the second man's sperm or indeed, if he were infertile, from a haploid nucleus taken from a cell elsewhere in his body. The baby would then be carried in a surrogate.

If such a prospect came about, along with the increase in active, elderly people, then future generations would be living in a society of parents of all ages. Indeed, it might become normal to have a child late in life, once one's career was over and there was more time and money to devote to the offspring. Children would grow up with the huge benefit of full-time parents, albeit elderly ones. But this new type of mothers and fathers would not be the equivalent of present-day grandparents; if they were physically much fitter, more mentally agile and more healthy than their present-day counterparts – as we have seen they may well be – then their age should not be a factor. However, there is a potential problem, not for the parent but for the child: whereas newborn cells divide some eighty to ninety times those from elderly organisms will do so only twenty to thirty times. The older the cell the shorter the telomere – the cap on chromosomes the loss of which allows chromosomes to stick to each other, resulting in eventual death of the cell.

It turns out that the first cloned mammal, the sheep Dolly, had shorter telomeres than usual. She died in 2003, having lived only half the usual lifespan of a sheep. But there remains some debate as to how much shortened telomeres influence longevity – and some recent data suggests that cells from cloned cattle actually seem younger than those from their conventionally conceived peers! Regardless of this, however, in April 2002 Ian Wilmut, co-creator of Dolly, reported that all cloned animals up to that date had suffered from genetic defects, including gigantism in cloned sheep and cattle and heart defects in pigs. As Wilmut warns, this research suggests a cloned human would be at huge risk of genetic defects – at least at present. If using DNA from somatic adult cells (not sperm or egg cells) means that the offspring from which they are produced have 'elderly' DNA, then their rate of ageing would be faster than that of those born of germ cells, an original sperm and egg. Then again, this situation could be offset by improved treatments for delaying cell-ageing, including treatment with the enzyme telomerase which will keep the telomeres long and healthy; in addition, everyone could be encouraged to freeze their eggs and sperm when in their biological prime.

A further complication, if reproduction and sex become increasingly distinct, will be the rise of surrogate mothers. A surrogate carries the biological child of the two commissioning parents, if the real mother is unable to carry the baby in her own womb; alternatively she may act as the genetic mother as well, following artificial insemination of the commissioning man's sperm, if the commissioning mother is infertile. There may be an emotional tussle if the surrogate mother does not wish, in the end, to surrender the baby. And there is arguably a potential psychological problem for a woman raising a child with whom she has no genetic allegiance, but the father does. Surrogacy in this form, then, may be less than ideal; in the future two further technical developments may offer alternatives, though not without the 'yuck factor' that often accompanies new technical discoveries and concepts.

One possibility, in cases where the baby is the genetic product of both commissioning parents, might be for a primate species other than human to act as surrogate. Already xenotransplantation, whereby a transplanted pig heart, for example, could save the life of an otherwise terminally ill coronary patient, is a real prospect; a surrogate womb is surely not that different, when it comes to either ethics or rationale. And if animals can help us out not just with heart problems but reproductive ones too, zoologist Robin Baker predicts, non-human surrogates could help men also – a man who had lost his testes, say, during cancer treatment. It would be possible to inject that man's stem cells, from which his sperm would be made, into the testes of several rats. The surrogate testes would then start to produce both rat and human sperm. These testes could then be grafted into the human scrotum, so that the man would be using surrogate rat testicular machinery to ejaculate his own sperm in the usual way. The only complication would be that he would be producing rat as well as human sperm, which may be a 'yuck factor' both for him and his partner. However, aside from an additional problem of possible allergic reaction to rat sperm, there would be no harm in the insemination of a human with rat sperm; in the unlikely event of it entering a human egg, that egg would die. By comparison to this scenario, or indeed to the heart xenotransplant, a surrogate primate womb seems perhaps more acceptable. Indeed the rat testes, or the pig heart, would be in your body for the rest of your life, whereas your offspring would be in an alien body for a mere nine months. Once it was born, no one would know any difference.

The other possibility would be to dispense with the squalor of biology altogether and use artificial wombs. This idea was actually first mooted by J. B. S. Haldane in 1923, in his prophetic paper

Daedalus, or, Science and the Future; a little later the notion was taken up by Huxley to form part of his *Brave New World*, in which babies are no longer born 'viviparously', from living parents. Scientists first explored this idea in 1969, when they managed to keep a sheep foetus alive for two days. More recently, in Japan in 1992 a goat foetus survived and was delivered after seventeen days in an artificial womb; but it was already 120 days old, three-quarters of the way to full term, before it was put in the artificial environment. The difficulty with developing an effective artificial womb that could sustain a human body from conception to full term is in simulating the highly complex and sophisticated workings of the placenta, which absorbs the goodness from the mother's blood, and dispatches waste.

A halfway measure might be to create an artificial womb lining, in which embryonic cells could be stimulated to grow, by means of appropriate combinations of drugs and hormones. The whole ensemble could then be transplanted into the mother. Such a treatment might be helpful to infertile women, and they would still undergo a conventional pregnancy. Assuming however that by the middle of this century, say, the huge technical hurdles are overcome, then artificial wombs might offer an attractive and ethical alternative to pregnancy.

One advantage would be that both parents could watch the child grow day by day through glass walls of the womb or via a computer-camera link; undoubtedly this would be a fascinating experience, at the very least, and the father, for the first time, would feel as involved and equally bonded to the child during the pregnancy as the mother. Now imagine the addition of augmented reality systems: different stages in the baby's development, along with changing statistics – the approximate number of brain cells, weight, heartbeat and blood pressure – all appear on the display. Obviously, with such a system, samples of amniotic fluid could be taken every day and would offer the paediatrician a much more accurate means of monitoring that all was well, and forewarn of any potential complications.

Aside from all the medical advantages for the well-being of the foetus, an arguable bonus would be that women could be increasingly active in the workplace if their babies were gestated in artificial wombs; they would also avoid the tribulations of normal pregnancy, with all the attendant problems of morning sickness, weight gain, fatigue, stretch marks, varicose veins, insomnia and other discomforts, as well as the increased risk of hypertension and diabetes. Of course, there is also the alarming prospect that different additives and supplements could easily be added to the equivalent of the amniotic fluid to create appropriately able or servile future citizens, as in Huxley's Central London Hatchery. But then, it is important to remember that such engineering could occur before implantation in any type of womb, artificial, surrogate or 'natural'. However, any queasiness – and there may be much justified objection to artificial wombs – should not include the potential for external intervention. Already there is no technical divide between screening embryos consisting of some eight cells after IVF, for elimination because of an unwanted trait, and screening them for implantation because of a desired characteristic, such as tissue-matching with a sick sibling. Let's compound the dilemma still further by actually changing or deleting a rogue gene.

We have seen earlier that the problem with gene therapy in adults is the difficulty of accessing the relevant malfunctioning cells of the body; these cells number trillions and are already realizing their malfunctioning genetic destiny. But if our intervention occurred at the eight-cell embryo stage, or even at the sperm or egg (germ cells) stage prior to IVF, then that problem would be solved. Every cell in the body thereafter (somatic cells) would contain the desired genetic profile. This germ-line engineering is already well established as a technique in research animals; an appropriate gene is changed so that the animal, usually a mouse, will then have a particular genetic defect – a transgenic model of a condition such as Alzheimer's disease or Huntington's Chorea, as used in the experiment exploring the importance of the environment on genetic destiny described earlier in this chapter. Similarly, germ-line engineering has the awesome potential of changing the human gene pool for ever. Not just the individual that grew out of the embryo but also their children in every generation to come would have been, effectively, genetically modified. This is why the technique is currently unconditionally illegal everywhere in the world.

It is but one small step, and far easier than engineering the 100 trillion cells in the adult body, to manipulate the DNA of those embryonic cells, or better still of the germinal cells, the sperm and egg from which they were produced. However, the problem with germ-line engineering has not only been that, unlike after somatic therapy, the change would be passed on for ever from one generation to the next but also that there could be considerable, and undesirable, interaction between the engineered genes and other genes on the chromosome.

But now both those problems might be about to evaporate. Over the last few years Andy Choo, from the Murdoch Children's Research Institute in Melbourne, and a separate team, John Harrington and Huntington Willard from Case Western Reserve University, have each pioneered auxiliary artificial chromosomes. If the technology lives up to its promise – and it already appears to work in mice – then far more genetic material could be introduced into a cell, with far less interference and with far more accuracy and ease of manipulation. And even the problem of engineering immortality could be overcome, by a pioneering new technology that could reverse the change from one generation to the next. The new gene would be combined with the gene for an enzyme (CRE) that could effectively obliterate it. This killer gene could in turn be switched off so that it was active only in sex cells, and even then only when a particular drug was taken. So, you could take a drug that eliminated exclusively from your sex cells the artificial chromosome with its germ-line engineered gene. You would still have the engineered trait, but your ensuing children would not. This reversibility would give the flexibility that has been the chief deterrent against legitimizing germ-line engineering, whilst at the same time allowing for much more effective gene therapy against gene-related diseases. Indeed, Gregory Stock foresees that the ability to change genes with each generation could be a positive advantage, analogous almost to the upgrading of software today:

Imagine that a future father gives his baby daughter chromosome 47, version 2.0, a top-of-the-line model with a dozen therapeutic gene modules. By the time she grows up and has a child of her own, she finds version 2.0 downright primitive. Her three-gene anticancer module pales beside the eight-gene cluster of the new version 5.9, which better regulates gene expression, targets additional cancers, and has fewer side effects. The anti-obesity module is pretty much the same in both versions, but 5.9 features a whopping nineteen antivirus modules instead of the four she has and an anti-aging module that can maintain juvenile hormone levels for an extra decade and retain immune function longer too. The daughter may be too sensible to opt for some of the more experimental modules for her son, but she cannot imagine giving him her antique chromosome and forcing him to take the drugs she uses to compensate for its shortcomings. As far as reverting to the pre-therapy, natural state of 23 chromosome pairs, well, only Luddites would do that to their kids.

So the door would seem to be open for prospective parents to eliminate single genes from a foetus – the 'designer child' – or to survey a portfolio of desirable genes in the hope of cherry-picking desirable traits – the 'virtual child'. Sinister though such eugenics might appear, Gregory Stock defends the idea, arguing that different people would value different traits in their offspring, thereby preserving diversity in society; at the same time if, say, the IQ of everyone was enhanced, then that would ensure that the playing field was actually more level than in our current very divergent society, not less. Moreover, assuming that germ-line engineering were safe and free of side effects, what parents would not want to help their children as best they could? How would such a strategy differ from paying for extra tuition, or sending them to the best school you can afford? The values and the much-hoped-for outcome would be the same, only the method would differ.

Two big problems that cannot be argued away easily, however, and which are not tractable to pure technology, are first the misapprehension about what gene manipulation might do, especially with respect to mental functions, and secondly an effective speciation between the haves and have-nots worldwide. The idea of designer and virtual children is predicated on a clear understanding of the relation between gene and mental trait. Yet we have seen that such a link is far from obvious. True, we can tackle an aberrant gene in the hope of alleviating a disorder, but it by no means follows that by adding a gene we can enhance a normal function; even if we could, we might well change a host of other brain functions as well, as many different types of proteins were made, all interacting in the individual brain in ways that could not have been predicted.

The second problem is the worldwide implication of genetic enhancement. Even if the process were cheap enough to be commonplace in the developed world, it is hard to see how rapidly developing countries such as those in sub-Saharan Africa could catch up with this technology, when the majority of their citizens are still without access to clean water and have yet to make a phone call. Gregory Stock argues that we need not fear a shrinking of our human gene pool: even if there were as many as 1 million genetically altered babies born per year that would still be only 1 per cent of births worldwide.

But surely there's the rub. Imagine a world where a minority from the developed countries have no congenital physical or mental defects, let alone that they are healthier, stronger, age more slowly and have a higher IQ, compared with those from poorer nations. So little would the two groups have in common, so different would be their interests, agendas and potential, that they would scarcely interact. Or perhaps a more real risk would be that they *would* interact, but in a situation where the first exploits the second.

So might we, ultimately, see a speciation, a divergence of the human race into two separate species, the 'enhanced' and the 'naturals'? Closer to home, if enhancement were a commodity just for the rich, for the stratum in society, say, that can currently channel income into private education, then might such segregation even occur within a single society, a return to the rigid class divide of old, only even more definitive? The non-negotiable hierarchy of *Brave New World* might become a reality.

But there is still a further step that Aldous Huxley didn't contemplate. There remains one final tweak to a world where we manipulate genes with such ease and sophistication: a 'synthetic' genome. Currently the concept of a 'synthetic gene' is the addition of a new gene to a string of existing genes. Here the innovation, compared to traditional, longer-term methods of breeding, is partly that the same gene can flourish and function when taken from one species and introduced into another. For example, a natural gene that some refer to as 'synthetic' because it is taken from an insect can make grapevines resistant to attack. These alien genes produce proteins that kill the relevant bacterium, and have now been successfully inserted into an important variety of vine.

But entire genes do not need to be transported: one spin-off of the genome project is the development of tools for manipulating individual components within a DNA molecule. Researchers can then construct any sequence of base pairs, the rungs in the ladder of DNA, and the ultimate goal is the precise, targeted placement of artificial gene segments. 'New' genes may originate from another species, as when a viral or bacterial gene becomes part of the human genome, or they may come about as the result of random mutations of the existing DNA structure. These mutations could be the alteration either of 'normal' gene structure or of gene sequence in the genome. One way in which a gene could mutate, or be made to mutate, would be the release from a radioactive element of a high-energy particle that could enter the cell, and collide with a strand of DNA. Such an encounter fractures DNA, and although internal cell-repair mechanisms can usually reassemble the strand in question, several base pairs end up out of order or missing entirely. In any event, however they are produced, these new genes, spliced into the existing genome, are 'synthetic', though naturally occurring.

The latest technology is challenging even that constraint. Already the known sequence of a particular bacterium genome has enabled the Institute for Genomic Research in Maryland to determine the number of genes needed for survival – about 50 per cent. These genes can now be synthesized and inserted into artificial membranes. It is a very real possibility that these 'cells' could then divide, effectively creating artificial life. If so, eventually one would not need human donors or a Gamete Marketing Board at all; the virtual child could be just that, made flesh. In the future we may be able to type out a wish-list sequence of base pairs, and an automated DNA synthesis machine will produce the relevant string of DNA to order; we would need to be able to produce base-pair sequences that code for completely unique compounds.

One of the many hazards is that, once again, there may be additional unforeseen consequences; for example, many non-protein compounds may be expressed by reactions similar to those used by RNA to make proteins. Ultimately it may be possible to trigger synthetic life forms capable of reproducing themselves and to use as 'factories' to produce the compounds we need. After all, it is the sequence of base pairs within DNA that makes genetic code; any change will change the blueprint, and hence the product.

As yet no one knows the size of the minimal genome for sustaining life. We humans have some 3 billion base pairs, whilst a virus, not taxed with the strain of an autonomous existence, has a mere 10,000 bases. To date, the smallest genome known – 6,000 bases – belongs to the bacterium *Mycoplasma genitalium*. It makes sense, therefore, to start with bugs made from 'artificial' genes; Clyde Hutchinson, of the Institute for Genomic Research, is convinced synthetic viruses will be available within a few years.

The problem has been previously that within a single chain of DNA there are hundreds of thousands of molecular rungs on the ladder – the base pairs; but until recently molecular biologists could join together only a hundred or so, at most. But now Glen Evans, Director of the Genome Science and Technology Center, University of Texas, has got around that problem, and devised a method to eliminate 'junk' DNA. As a result, he has mapped out the way DNA is configured to create 'synthetic organism one' (SO1), a microbe. 'SO1 will have no specific function but once it is alive we can customize it. We can go back to the computer and change a gene and create other new forms of life by simply pressing a button.'

One application of this new technology would be the appearance of designer bugs for infecting target tissues such as tumours and then killing them. Or we could have our guts infected to produce Vitamin C. But just imagine the perils, intentional or accidental, of infecting humans and wildlife. If that is not awesome enough, imagine what would happen if SO1 could feed and reproduce, and thereby exist independently. Scientists like Hutchinson and Evans would have, in the eyes of some, lived up to the Dr Frankenstein image of scientists by creating new life. By manipulating and even creating genes future generations will lead very different lives from ours, and indeed they will view life itself in a very different way. Even taking the highly indirect relation of mental trait to genetic provenance into account, it seems likely that dysfunctions such as depression or schizophrenia will be much rarer, not only because of screening and intervention at the level of the gene but also because daily life will be more homogeneous, further removed from reality, less left to chance events 'out there', more rooted in a cyber-existence.

Yet a gene-driven reduction in mental anguish, as well as the almost certain alleviation of physical suffering, might mean that human nature will be sanitized. It is of course an age-old debate whether suffering ennobles and enables human beings to reach their true potential by coming to terms with the vicissitudes of life. What if that life was no longer highly individual? We have seen that not only might generations to come have a more limited and spruced-up gene pool but also that our manipulation of those genes could eventually break down the traditional stages of our life narrative. Soon there will be little reason to have more than one child with the same partner, and indeed each child could lay claim, in theory, to a variety of parents: the genetic donors, the egg donor, the surrogate mother and the parents who bring him or her up. Each of these claims to parenthood of one child could come from different individuals.

Clearly, our attitude to life, and to living, could well be transformed. If everyone is healthy and mentally agile into old age, if we all have a homogeneous lifestyle, passively receiving incoming sensory stimulation, if sex and reproduction are utterly segregated, if anyone can be a parent at any age, or indeed if parents as such can be abolished by a combination of artificial wombs, IVF, and even artificial genes, then all the milestones that mark out one's life narrative will be removed: being a child in a nuclear family, being a parent, being a grandparent, coping with the unexpected, with illness and with ageing. We may not ever be successfully enhanced to be super-clever or super-witty or good at cooking, as some might imagine, but in the future we may all adhere to a physical and mental norm, not just through the direct manipulation of genes but also through a new type of life made possible by such intervention, a life where proactive individuality, an ego, is less conspicuous, less used, less abused and less needed.

Some, such as Francis Fukuyama or the psychologist Steven Pinker, argue that human nature can survive changes in the environment, that it is an irreducible 'factor X' wired into our brains and bodies that makes us so special, and so different from all other species. But our bodies and brains are only composed of genes, the proteins they make, and other molecules that those proteins make, which in turn make cells. This shifting landscape of chemicals will be influenced dramatically in the future by gene manipulation – not to mention manipulations of the environment. In any case, we have seen that human nature is inextricably caught in an endless dialogue with human nurture, so let's now turn to the prospects for 21st-century child-raising – education.

6

Education: What will we need to learn?

Education is currently in crisis. As every year marks record successes in national exams so the protestations about dumbing-down become more shrill, whilst universities remain remote and expensive enclaves of an elite. Teachers are demoralized and parents angry and anxious. Given the existing culture of core curricula, pressure, audits, consultations and experimental new ideas, it is small wonder that there is no consensus on the bigger picture: what we should be teaching the next generation to equip them for citizenship in the mid 21st century, and beyond.

The large-scale changes in our lifestyle that might become the norm before too many decades have passed raise fundamental questions about the point of education as we know it, and most importantly about the type of mindset that 21st-century education will create. If the environment is about to change so radically then so will our minds; neuroscience and neurology are offering a wealth of examples which all illustrate a basic yet exciting notion. The human brain reflects, in its physical form and function, personal experiences with supreme fidelity.

Soon after his birth it was clear that there was something wrong with Luke Johnson. The little boy couldn't move his right arm or leg: he had been the victim of a stroke just as he was about to enter the world. But over the next two years the paralysis slowly, seemingly miraculously, receded. Luke now has completely normal movements – and he is far from unique. A staggering 70 per cent or so of newborn babies who have suffered disabling strokes in the peri-natal period regain mobility. The brain, we now know, is able to 'rewire' itself. It is this 'wiring', the connections between our brain cells, that makes each of us the individuals we are. However, the electronic metaphor of fixed hard-wired circuits is not really appropriate, since it misses the critical point: our connections in the brain are constantly changing, adapting to our experiences as we interact with the world and live out our personal set of experiences. So just how might the experience of 21st-century life leave its mark on the brains of upcoming generations?

When a baby is born he or she has a far greater density of connections (synapses) between neurons than an adult does. However, neuroscientists cannot easily study, on a large scale, precisely how these connections configure in different regions of the human brain from one person to the next: the distressing attitudes of certain cavalier and unscrupulous pathologists in the past, combined with a conviction that human body parts are an integral part of the deceased, now mean that human brains are usually destined not for laboratories but for funerals; the valuable clues about mental function that they might contain are therefore locked away inside them and lost for ever.

Yet from the limited numbers of autopsies of the human brain that have been possible, neuropathologists have discovered that synapses in the outer layer of the brain (cortex) relating to vision peak at about ten months; after that, the density slowly declines until it stabilizes at about ten years of age. But in another region towards the front of the brain, the prefrontal cortex, the formation of connections (synaptogenesis) starts conspicuously later and the subsequent pruning of those contacts takes longer than in the visual regions; in this case the density of connections starts declining from mid adolescence and reaches a plateau only at eighteen years old.

How do these developments in the physical brain relate to the development of mental abilities? Everyone recognizes that the first few years of life are critical for the acquisition of certain capacities and skills. The most obvious deduction, therefore, would be that the number of synapses at any one stage of development is linked to the sudden appearance of some new ability. But the problem is this: we now know that these skills carry on improving, even after the densities of connections between neurons dwindle to adult levels. So what really counts, ultimately, in determining how information gains long-term access to our neurons? It turns out that, rather than brute number, the *pattern* of connections must be all important. Nature provides a plethora of synapses over the first few years of life, which grow as the brain does; these connections are long enough to connect up vast tracts of brain terrain. 'Sculpturing' (a clichéd but highly appropriate label) then occurs, whereby supernumerary contacts disappear – and just as a statue emerges from a block of stone so a unique brain takes shape. But unlike a stone statue, the individual pattern of connections that make up your brain remains highly dynamic. As you experience each moment of life, each event will exaggerate or blur some aspect of the overall design within your head.

In 1981 vision experts David Hubel and Torsten Wiesel won a Nobel Prize for an astonishing discovery: in the developing brain, it turned out, there are particular windows of time, 'critical periods', in which certain large-scale wiring occurs. One particularly moving example of critical periods at work in real life is the story of a small boy who, aged six, presented as a medical mystery: he was blind in one eye, even though the eye seemed perfect. Only after extensive questioning of the parents did it emerge that he had had a minor infection when he was less than a year old. In itself the condition was trivial but the treatment had involved bandaging the eye for several weeks, weeks that corresponded to the 'critical period' for the eye to establish appropriate contact with the brain. As a result, the unclaimed neuronal territory was invaded instead by connections from the working eye, so that when the bandage was removed there was no brain space left for the previously infected eye: it was useless therefore, and the boy was blind in that eye for the rest of his life.

Usually, however, the windows of time are not as rigid nor are the effects of 'missing' them as irreversible. For example, there is a naturally occurring equivalent of the ocular deprivation imposed by the bandage. Sometimes babies are born with cataracts. Surgery can often help enormously and, of most relevance here, is often more effective when there has been a cataract in *both* eyes. If neither eye is working normally, and thus remains unstimulated by the ongoing act of seeing, then the brain territory will go unclaimed: there are no connections from a working eye to invade it. But after surgery, once the eye on each side is able to function, each will hook up with the relevant part of the brain. These observations show that some recovery of function is possible, even though certain time frames are clearly very important.

The notion of key periods for development of a basic brain function like vision has led some educationalists to speculate that such windows of opportunity might exist for more rarefied activities, such as reading and arithmetic. As yet such a question is hard to address; so many different factors must contribute to literacy and numeracy that it would be hard to dissect, across a wide range of diverse individuals, the isolated single factor of age. Moreover, we are very likely to learn in different ways as we grow older. A young child will absorb any incoming information with scant sales resistance (remember the Jesuits' promise, 'Give me a child until he is seven, and I will give you the man'). However, the older we become the more any experiences, including formal teaching, will be measured up and evaluated by the less accepting, maturing mind.

These checks and balances that make up an individual's mindset have their root in the connections between brain cells. For the most part these connections converge onto a zone on the target cell named after the Greek for tree, 'dendrites'. Dendrites are so called by virtue of their physical shape, which does indeed resemble the branches of a tree, and, like trees, some neurons have more extensive branches than others. The greater the ramification of dendrites the more readily a cell will be able to receive signals from incoming neurons. The basis of brain growth is not the increase in bulk numbers of neurons themselves but is primarily a story of the proliferation of the dendrites.

These dendrites will configure in a way that reflects what has happened to you. A pivotal, classic experiment in rats shows that the post-natal environment exerts a massive influence in determining how widespread this ramification of connections will be. Scientists compared the effects of an 'enriched' environment, complete with rat toys – ladders, exercise wheels and the like – with the 'normal' lot of a lab rat – a warm home cage with food and water but little else. Post-mortem examination did indeed reveal that the brain cells of rats who had experienced enrichment had more extensive branches than their more typically housed counterparts. Although this is clear evidence of how the stimulation of everyday life can make a difference to the brain, neuroscientists nowadays are quick to point out that for a 'normal' feral rat 'normality' would be something more like the enriched environment, whilst the lab situation, sadly, is more truly akin to deprivation. Yet the findings are surely all the more chilling as a consequence.

In 1999 the psychologist Thomas G. O'Connor and his colleagues explored what the implications of environmental deprivation might be for the human brain by studying Romanian babies who had spent the first years of their lives in the infamous national orphanages where there was very little varied sensory or social stimulation. Perhaps not surprisingly, the researchers found that the children were more likely to have delayed walking and talking skills, as well as impaired social, emotional and cognitive development. The opposite, accelerated skill acquisition, is the aim of the hyper-stimulation in the hothousing of infants – the frantic attempts by some parents to provide intensive stimulation for their children when very young, in the hope that they will excel in later life.

Whether this strategy is necessarily successful is under considerable debate. There has been some concern that hothousing may have a negative effect – resulting in low self-esteem, a deep sense of failure and a tendency to underachieve. Further, even when the method has the desired academic result, children educated more intensively than their contemporaries frequently experience emotional and social difficulties in later life. 'I can't understand why people enter their young children for exams, unless it's parental pride,' says Professor Joan Freeman, author of *Gifted Children Grown Up*.

The clear adaptability (plasticity) of brain cells, with scant or extensive branches, reflects the whole gamut of minimal and maximal stimulation that can come the way of the human brain, and although there may be sensitive periods when the changes are most dramatic the burgeoning and withering of dendrites will continue to occur into, and for the duration of, your adult life. As your neuronal connections grow, shaped by your particular experiences, so the dialogue between your brain and the outside world becomes more two-way. Instead of seeing life in abstract, sensory terms – how sweet, how cold, how loud or how soft – those sensations coalesce into people and objects. As these people or objects feature repeatedly in your different experiences a growing number of associations will form around them via your growing dendrites; they will be of increasing significance, will 'mean' more. The individualization of the brain will increase as vast ranges of brain-cell circuits configure in extent and power according to the particular types of input they have, that incessant and complex assault on the senses that makes up your daily existence. This forging of new connections, which has a direct basis in the connections between neurons, is surely the essence of learning.

A few years ago, one fascinating example captured the imagination of the media: the findings, based on brain scans, revealed that a certain region in the brains of London taxi-drivers was physically larger than in non-taxi-driving individuals of a comparable age. Since the area in question (the hippocampus) is related to memory functions and since London taxi-drivers have impressive memories, needing as they do to learn the lay-out and names of all the streets of London by heart, here surely is a clear demonstration of how the brain, even in adults, responds to stimulation.

In another report, brain scans have revealed that in highly skilled musicians there is an increase of 25 per cent in the size of a key part of the brain related to hearing (auditory cortex) compared with people who have never played an instrument. And more telling still is the observation that this increase matches up with the age at which the individuals began to practise rather than when they achieved pro-ficiency. The critical issue, it seems, is the activity itself of practising music, not how good you are.

A further experiment, again with adult humans, proves that you do not have to volunteer for a change in career nor practise at music for long to change the size of functional areas of your brain; instead, you can enrol in an investigation of the effects of five days of piano playing for two hours each day. In such a study the subjects were all non-piano-players and were divided into three groups. Group 1 were merely exposed to a piano and left to play around with it as they wished; Group 2 started to practise five-finger exercises, whilst Group 3 had simply to imagine they were playing the exercises. Perhaps not surprisingly, the area of the brain relating to the movement of digits dramatically expanded in Group 2 compared to their uninstructed colleagues in Group 1. However, the truly amazing result was that Group 3, those who had engaged in non-physical mental practice, had brain changes almost as impressive as those who had acted out what they were only rehearsing in their minds. Apart from discrediting once and for all the old dualism of mental versus physical, of mind versus brain, such experiments surely ram home the point that what you do is reflected in the fine architecture of your brain, and that a particular configuration of your brain cells will enable you to perform a particular skill with ever-increasing facility.

But such exaggerated studies on the effects of experiences are only the tip of the neurological iceberg. As for the rest of the body, the more any particular part of the brain is exercised the more effective it will become. This efficacy, in brain terms, means the proliferation of dendrites and hence the appropriation of more brain territory. On a much more subtle scale everything you do, and everything that happens to you, will leave its mark, literally, on your brain. The human brain, after all, is very good at learning; our ability to adapt to our environment, to learn from experience, distinguishes us from all other primates, even chimps. Our singularly human brains have enabled us to occupy more ecological niches than any other species on the planet; the capacity of our neuronal connections for adaptation has freed us from the genetic tyranny of a generic instinct. Different cultures geographically separate in space, like generations separate in time, differ so much from each other because the respective brains have been exposed to such different influences.

As a developing individual you see the world in terms of what has gone on before, in your unique trajectory, and slowly transform from an undiscriminating data-sponge to an information cherry-picker. The process of assimilating information may not now occur with the same unconditional and effortless facility as when we were young enough for the Jesuits, but understanding – seeing one thing in terms of another – will be increasingly possible. It is this unique personalization of brain-cell circuitry that, in my view, is the physical equivalent of 'the mind'. Easy to see then how the minds of our cave-dwelling ancestors would be different from our own; easy to see, also, how technology has accelerated as each generation has learnt so effectively from those preceding it, and we have been able to stack up our own discoveries on an existing body of knowledge. In order to explore the extent and manner in which the new technologies will shape the young minds of this century we need to identify the key factors in the learning process.

The idea that any stimulation is good, simply because it's stimulation, must be simplistic. In any case, we cannot assume that indiscriminate stimulation of any one type is all there is to learning. If a new skill such as taxi-driving in London or playing the piano can enlarge a brain area, it can surely do so only at the expense of some other skill: what might we become less good at? And since we now know that any one brain region will participate in more than one net behaviour how far can we generalize? Would the taxi-drivers with an enlarged hippocampus also be better at the host of other functions with which the hippocampus has been linked?

Another very basic factor in determining how effectively the brain learns could be sleep. In rats, at least, periods of learning are associated with increased dream REM (Rapid Eye Movement) sleep. And REM deprivation impairs rat memory. More persuasive still, Dr Pierre Maquet and his team showed a few years ago that in humans, during sleep, some brain areas are more active in trained than in non-trained subjects, whilst the effects of that training are improved still further the next day after the opportunity for a period of dreaming.

Dreaming has been a source of fascination for thinkers down the ages and in the present day continues to be an extensively researched phenomenon among neuroscientists; but the purpose of dreaming is still open to conjecture, as is the precise course of events that unfold in the brain as we enter that eerie, utterly subjective and irrational inner world. Many believe that dreams help us to consolidate the thrills and spills of each waking day, yet this type of 'explanation' could just as easily be an effect of dreaming rather than its cause. Dreaming could simply be a form of consciousness not driven by the normal sensory inputs, so that the net experience is very different and far less constrained by the 'reality' of the outside world. If so, the main trigger would be the residual activity of the brain, which would happen to reflect most recent events. Memory consolidation would therefore be a corollary of dreaming but not the essential driving force. After all, small babies dream, even in the womb and even more than adults, and they have very few life crises to resolve! So perhaps dreaming is simply that residual brain activity itself, undisturbed as it is by inputs from the senses, and as such is all important in learning. As the connections between neurons in the brain, at any age, rehearse over and over their electrochemical sequences so they become ever more effective and efficient.

In addition to sleep, further factors are being suggested as influences on how readily the brain learns. One experiment in particular, first performed in 1993, has generated huge controversy and speculation. Volunteer subjects had to work out what a paper would look like after being folded and cut in a certain way, like a paper doily. After the test, one group sat in silence for ten minutes, a second group listened to a Mozart piano sonata whilst a third heard an audiotaped story or repetitive music. All three groups then took the test again: the 'Mozart' group accurately predicted 62 per cent more shapes on this second test, whilst the silent group improved by a puny 14 per cent and the repetitive music/story group by only 11 per cent. Although such clear-cut and dramatic results have defied replication so far and many are still very sceptical, Lois Hetland, of the Harvard Graduate School of Education, extended the study to 1014 subjects. She found that Mozart listeners out-performed other groups more often than could be explained by chance.

In fact, a completely different type of experiment seems to support the notion that for some reason Mozart may be good for the brain: it turns out that rats raised with the music of Mozart run mazes faster and more accurately than other rats. Rodents are hardly renowned for their appreciation of the great composers; clearly, then, whatever the effect is it has little to do with musical sophistication, or even with the notion that listening to music puts you in such a good mood, or so arouses you, that your performance improves too.

One clue to what might be happening comes from the work of Gordon Shaw, from the University of California at Irvine: amazingly enough, the electrical discharges within networks of neurons sound like music when expressed acoustically. Could the patterns in music conversely drive the formation of networks of neurons, which are consequently primed for more efficient mental function?

John Hughes, a neurologist from the University of Illinois Medical Center, has shown that a critical factor in the intellect-enhancing effect is how often musical volume rises and falls in surges of ten seconds or longer. The music of Mozart scores two to three times higher than minimalist music or pop tunes in this respect. It seems that the regular, repeating sequences of twenty to thirty seconds may fit best with brain-wave patterns of thirty-second cycles. Only specific music, then, will stimulate the brain in the right way: indeed, brain scans of volunteers listening to Beethoven's *für Elise* and 1930s popular tunes show that only the auditory part of the outer brain layer (cortex) is activated by these very different types of music – but Mozart makes the cortex light up all over!

These fascinating findings raise far more questions, for neuroscientists and educationalists alike, than they answer: not least we need to know how outside stimuli can train brain circuits and how acoustic priming, somehow, improves our ability to think. The possibility that activities remote from formal education such as dreaming and listening to Mozart can enhance learning ability may well feature in strategies for education in the future. Imagine 'covert priming' sessions as both a pre-school norm and a commonplace warm-up exercise before each formal lesson.

Further into the future still a new phenomenon in the classroom might be even more precise and direct exploitation of inherent brain mechanisms. We have seen already that awesome strides are being made in brain-imaging techniques; quite soon, perhaps, the time frames over which we can monitor brain events will be commensurate with the split-second real time over which neurons operate. Yet so far no one has really given much thought to improving such windows onto the living brain in space as well as in time. Surely one day some bright technocrat might come up with a way of monitoring the plasticity of the brain – the atrophy and growth of dendrites – on the fine spatial scale of individual neurons in the living, conscious brain.

Let's imagine the implications, if such high-resolution combination of time and space frames were possible. The subject of a brain scan currently has to journey to the lab or hospital in order to be buried within the huge cylinder that houses the colossal magnets needed. Yet this might be comparable with the first computers, which occupied whole rooms and had only a fraction of the capabilities of a present-day palmtop. By analogy perhaps one day the expensive, technically capricious and clunky imaging equipment we have now might be replaced by an elegant helmet. In this way it would be possible to monitor the formation and disbanding of dynamic neuronal networks in normal environments such as the classroom, and indeed to watch the brain processes that accompany learning. The teacher could then observe a console of screens and see how effectively a child was primed by, say, Mozart before commencing formal instruction.

Let's speculate still further. Once patterns of connectivity could be precisely documented, localized and matched up with certain types of learning, then it would be just a small step from monitoring to manipulation. Perhaps non-invasive radio-stimulation, transmitted via the helmet to certain neuronal assemblies, would drive the pattern of connections into the desired configuration. Lest anyone think that this whole notion is completely absurd, please note that the neuroscientist Dr Michael Persinger is already stimulating the brains of human subjects in just this way, admittedly without anything near the anatomical precision that would be required to manipulate specific neuronal networks. The focus of his experiments has been to stimulate the mind into having a 'religious experience' – although exactly what the experience might be varies considerably from person to person, and the area to be stimulated cannot be accurately targetted.

Yet already a greater finesse in brain location can be achieved with a 'Gamma Knife' – a device that uses ionizing radiation to allow neuro-surgeons to operate on abnormal areas of the brain without making an incision. This technique might not yet be able to stimulate select neuronal groups but is still capable of destroying minute amounts of target brain tissue or tumour, with great precision. Perhaps, then, in the future some combination of these two techniques – a stimulation confined to highly specific groups of brain cells – could realize the most direct method of 'teaching' of all, where no active learning was required . . .

Yet the invasive programming of the brain with facts, like the non-invasive strategy of hothousing, does not necessarily deliver the desired result. Moreover, in the cyber-age facts will be so accessible that there will be little need to internalize them. And even when a fact is learnt, that doesn't guarantee understanding. Facts on their own have always been intrinsically meaningless. A fact acquires a meaning only once it is associated, linked to another fact. For example, when my brother was very small I taught him the famous soliloquy from *Macbeth*, where the eponymous hero despairs: 'Tomorrow, and tomorrow, and tomorrow, / Creeps in this petty pace from day to day /

To the last syllable of recorded time; / And all our yesterdays have lighted fools / The way to dusty death . . .' Graham, at three years of age, could recite the whole piece word perfect but he didn't understand it. After all, how could a toddler grasp the metaphor of 'petty pace' and understand the meaning of 'dusty death'? He would have had to acquire a huge database of prior information first, and then been sufficiently adept at linking those facts – been educated – to make the multilayered cross-references that we recognize as 'understanding'. Then again, wholesale yet precise brain stimulation would be all the more sinister if it circumvented both these drawbacks by driving connections between isolated facts, thereby inducing an automatic understanding, a complete programmed mindset.

But back to the foreseeable future, where children will still be relying on the more haphazard and indirect stimulation of everyday life to configure their brains, and hence shape their minds. Another factor that is and will continue to be highly relevant is learning-by-doing, as anyone would testify who ever tried to learn to drive by watching someone else. Children's main sensory and cognitive learning achievements come from their own experiences in the course of activities such as play, exploration, everyday talk and social interaction with peers and siblings. Inevitably then, early experiences constitute an important factor in how well children assimilate information in more formal learning situations later on. The negative effects of low-income backgrounds are now well known, publicized by a programme in the USA designed to offset the disadvantages.

This scheme, Head Start, aimed to provide pre-school compensation for infants not, perhaps, as deprived as Romanian orphans but socio-economically disadvantaged nonetheless; the experimental curriculum aimed to develop a whole range of physical and mental skills, including sharing and counting objects, fitting objects together, anticipating and remembering sequences of events, role playing, imitation, recognizing objects, playing simple musical instruments and talking with others.

The Head Start programme pays. If you consider the financial cost to society of juvenile delinquency, remedial education and income support that could not only be saved but also further offset by taxes from higher paid jobs, then it's not surprising that Head Start returns impressively sevenfold on each dollar invested. Indeed, it is particularly telling to compare children that had been in this programme, those that had been on a regime of direct instruction and those who had simply played at nursery school: the first two groups showed a higher IQ than the nursery-school children at school entry, whilst at fifteen years old the Head Start and nursery-school children were committing 50 per cent fewer delinquent acts than those who had had the intellectual advantages of direct instruction without the benefits of a more socialized experience; in fact, by twenty-three years of age those who had had only formal teaching pre-school were at a distinct disadvantage on a range of personal and emotional measures. The Head Start programme therefore offered the best of both worlds, an intellectual as well as a social advantage.

These results show that at four or five years old children have still not fully developed their social and cognitive skills to maximize the benefits of learning from formal instruction. For small children, then, the best option is self-initiated play and exploration, not formal, academic study. Ideal programmes involve a high degree of parental involvement, with the space and time for children to play and discover things for themselves. Inevitably, therefore, we need to look at what is currently the most immediate and most powerful influence on young minds: the family unit.

Nowadays, only 17 per cent of American families conform to the traditional profile: a stay-at-home mum, a breadwinning dad plus two kids. Families, in the developed world at least, are becoming much more diverse or merged. More than a third of all children born today will live in some kind of step-family household before they reach eighteen. An unsurprising prediction is for shorter, later, less sacrosanct marriages with easier, quicker, less traumatic divorces; eventually, perhaps, the requisite legal procedures could take place merely by video-conferencing or on the web. Divorce itself is down to 43 per cent from a peak of 51 per cent of all American marriages in 1981, but this decrease could be due to marriage having become less popular in the first place. Accordingly, cohabitation is way up from 500,000 unmarried couples in 1970 to 3.7 million in 1995, with almost half of all children in the USA living at some stage in a cohabiting family. This trend towards increased cohabitation cuts across every age, ethnic and economic group.

Other factors contributing to the increasing heterogeneity of the family will include the swelling number of stay-at-home fathers, a number grown by 25 per cent to 2.1 million in the USA over the last three years. There will also be more couples sharing homes, as a result of the growth in house prices and the needs of economic immigrants. Then there is the issue of monogamy. Hugh Hefner, the philandering founder of the magazine *Playboy*, concedes grudgingly that it may well still be a 'viable choice' in the future. A variation on traditional monogamy is serial monogamy, in keeping with the increased fluidity of the family unit, which is indeed on the increase. And to stretch the concept of sexual fidelity beyond all practical recognition or relevance, writer Adam Phillips suggests that it is the *values* of monogamy that will actually last into the future – loyalty, fidelity, long-term affection – but not necessarily the idea of having sex with only one partner.

Not only will sexual partnerships continue to become more and more fluid but the once robust structure of generations will also buckle. We have seen that the quality of life for older people in this century is set to improve massively, but it is not clear how they would relate to the on-going nuclear family. Older people could well be physically and mentally fitter, perhaps even become parents themselves late in life. Perhaps they will exist as independent, isolated individuals living apart from the nuclear unit of their children and grandchildren; or perhaps the family itself will disintegrate and everyone will be living on their own as soon as they can, as single person units; or a third possibility is that the notions of family and extended family will become so diverse and vague that effectively any relationship of any type could be subsumed under that label. In any event, an increase in the elderly population will be just one factor in a world undergoing colossal social upheavals.

But remember the more distant future holds the prospect of a more homogenized individual – one free from generational stereotyping, and more similar to others not only in gene pool and physical fitness but also in mindset and outlook. Add to this prospect an increased tendency to be the passive recipient of a cyber-world and it seems increasingly likely that our successors will be far more interchangeable. There may well be a greater shift of personnel within a family unit but paradoxically less actual diversity in lifestyle or in attitudes and relations within each family unit. An interesting question, therefore, is not so much to what extent or frequency new step-parents, grand-parents, wives and husbands might supplant the old but the degree to which such swapping around of highly similar individuals ends up being particularly relevant.

'One of the major problems which has emerged at the end of this [20th] century is the large number of influences on children other than parents. Parents will always be important, but there is real compe-tition,' cautions Sylvia Rimm, author of *Smart Parenting*. Already children are exposed to beepers, cell phones, video arcades, 24-hour movie channels, internet websites, email, chat rooms, online shopping and virtual reality. There is a very real likelihood that parental influence will shrivel. So whilst it is important to place a premium on stimulation of the brain, it could well be that the source of that stimulation might be changing from the family to the computer. And if so, then the children of the future might have a very different outlook as they rely increasingly on the cyber-world.

Now cyber-technology is starting to merge with that most enduring of childhood influences – toys. In 1999 toy-industry sales reached $71 billion. Toys are clearly big business, and with software prices dropping it is inevitable that we will see more and more 'smart' toys. Already, the last few decades have seen an increasing sophistication in the degree to which toys are interactive. Things have really moved on from the 'speaking' doll with a small gramophone in her stomach that I received one Christmas in the 1950s. Now remote-controlled, multi-functional toy cars can be customized through a CD or the internet, whilst toy trains will stop or accelerate under voice control. Meanwhile My Real Baby becomes more 'independent' over time; subtle software progressively gives the impression that the doll is gradually initiating her (always her) own actions and preferences. Another cyber-hearted infant, My Dream Baby, actually develops physically by 'growing' four inches, and progresses from crawling to toddling; not just physical but also intellectual development is apparent, as the doll appears to 'learn' new words using voice recognition. This new toy cyber-world is not just inhabited by surrogate babies but by pets as well. For example, Robokitty, all of ten inches long, has video-camera eyes, stereo-microphones for ears and speakers for meowing; thanks to touch sensors in appropriate places, kitty can even purr when touched.

The IT expert Mark Pesce explores in his book *The Playful World* how toys have always performed the role of introducing children to the 'complex universe of human culture' as they interact with them. But nowadays, unlike in the past, the physical world is becoming exquisitely interactive, and therefore malleable. We have just seen that interaction is a vital component of learning; indeed it is widely acknowledged that for people of all ages, especially the elderly, it is important for mental health to maximize the opportunities for control of the objects around them. So on the whole, the increased potential for a child, or anyone, to be able to manipulate his or her environment would surely be a good thing.

On the other hand, I wonder just how much life in a predominantly transient world could impact negatively on the young mind. Children seem to need a sense of routine, a consistent set of faces, values and rules. If you realize early on that anything and everything around you can change, be it in appearance or context, at the press of a button or even at a voice command, then how might you start to conceive 'reality'? Imagine you can alter everything around you; you may consequently have a very shaky concept of how you relate to this inconstant reality – your own sense of identity might also waver, or never even establish itself.

High-tech toys that may eventually make such a big difference to young thinking really started to take off in the last few years of the 1990s, with Tamagotchi, who featured briefly in Chapter 2. Then the technology was sufficiently modest to amount merely to a virtual 2D pet inside a small plastic shell. However Furby, born in October 1998, was the first real electronic furry toy; not only was it reactive and verbal but it existed in 3D. No wonder initial demand outstripped supply by four to one. Furby has a righting mechanism, the ability to 'fall asleep' and 'wake up' thanks to light sensors, plus a microphone in the ear. All the underlying software constitutes 'brainpower' merely one ten-billionth of our own, yet the toy still seems alive because it plays on our instinctive desire to anthropomorphize. The facial expressions with which Furby can react are easily interpreted, especially by its young owner, as wonder, anger, sleepiness and joy. And to add to the impression that Furby is really one of us the software can slowly release a repertoire of sayings that replicates the linguistic ability of a child in its first five years. A crucial feature is that, like Tamagotchi, Furby has needs, for example, the need for 'nourishment'. However, unlike the two-dimensional forerunner from Japan, Furby has a real tongue that the carer must press. If this activity is insufficiently frequent – if it is 'underfed' – then Furby will sneeze. Moreover, there is the illusion of 'learning', as different behaviour patterns that mimic human relations are gradually revealed.

In the future toys will routinely be as truly 'ignorant' as a newborn, but slowly 'learn' from experience – as we saw the next generation of working robots might do. This new generation of toy infants will assimilate, as a result of feedback, the many expressions of the human face in parallel with their young owners. Moreover, these synthetic infants will follow their human controllers, explore and start to recognize objects in parallel with them. Hence the toy will serve as a mirror to the child's own development, acting as a kind of friend but always one who is not an autonomous equal, more of a servile sounding board.

Another change might be that although children of the future have a less defined sense of self they are used nonetheless to compliance with their will. Encounters with other children could end up troublesome, and therefore children may increasingly shun each other in favour of a more biddable cyber-society. Although at the moment children can apparently readily distinguish between real love and 'Furby love' we should not be so complacent as to expect that this might always be so. And as adults interact increasingly with non-biological beings so children in the future may grow up with a very one-sided view of relationships.

How will the child relate to the outside world on a global level? This will be an even more invasive and pervasive part of education than learning from cyber-toys. The implications for education are enormous when it comes to the difference the web is making to our lives, as a tidal wave of facts deluges us from the screen. At the most physical level our literary activities will become ever more computerized. The printed version of the *Encyclopaedia Britannica* costs, at the time of writing, some £1,000. A CD-ROM replacement is yours for just £50, and will be replaced by the net any day now, as soon as we are all confident that nothing has been omitted. All non-fiction books are in effect already competing with a free online resource, which is growing every day. The cost of all information is rapidly falling to zero. There will be instant access to all facts, for everyone. But what will future generations do with those facts?

A child born at the dawn of this century will never know a world without the web, but most significantly it will increasingly become a web that reacts. This highly accessible and interactive dialogue that younger people are already taking for granted is perhaps one of the biggest factors that will drive a wedge between the generations in the next couple of decades or so. For those, like me, on the 20th-century side of the divide, the 'freedom' of the web will, as Mark Pesce predicts, be 'chaotic', 'disorientating' and 'discomforting'. But for the young of the 21st century there will be an unquestioned assumption, a confidence, of access to instant information. Yet the issue goes deeper still: it is not simply about what we do or do not need to learn, but how we think.

A recent piece in *Time* drew a clear distinction between we who grew up in the second half of the 20th century, 'People of the Book', and the new generation, 'People of the Screen'. We People of the Book work within a culture of newspapers, law, offices of regulation and rules of finance. Most significantly, the foundation of this culture is captured in texts. From an American perspective it would be East-Coast based. By contrast, the upcoming People of the Screen are culturally more West-Coast based, working with the TV, computer, telephone and film. The journalist Kevin Kelly writes:

Screen culture is a world of constant flux, of endless sound bites, quick cuts and half-baked ideas. It is a flow of gossip tidbits, news headlines and floating first impressions. Notions don't stand alone but are massively interlinked to everything else; truth is not delivered by authors and authorities but is assembled by the audience.

The People of the Book, according to Kelly, fear that logic will give place to code, that reading and writing will die. As far back as the 1960s the futurist writer Ted Nelson made a prediction along similar lines, declaring his vision of a world in which 'no longer would we be stuck with linear text, but we would create whole new gardens of interconnected text and graphic for the user to explore'.

Again, then, there may be a reduction not just in the constancy of the physical world, but also of unambiguous and accepted bodies of knowledge. All current technology points in this direction. Voice-activated systems have not turned out to be quite as easy to develop as was originally thought, but no one denies that sooner or later they will be commonplace. And if we also have technically simpler oral computers, then it is hard to see why the typical child of the near future will need to be literate. After all, we speak far faster than we write, some hundred words a minute. In the future, speaking into your computer at 6,000 words an hour, you would be able to write a complete novel (were the concept still to be valid) in less than twenty hours.

Yet we can take the possibilities further. Your finished novel could be available to the reader in multiple versions. Moreover, if they keep it in its electronic medium, accessing it on their screen rather than downloading it onto paper, they could introduce not only hypertexting references but visual material too. Then the story could be interactive: the 'reader' could not only select the course of the plot but also impose their own faces and those of their family in the visuals of the main characters. In the transition period, whilst we still use words on the screen, we might end up with TV that we read and books that we watch! Again it is possible, if not likely, that the boundaries will blur between reader and writer, between fact and fiction, with the consumer in control, a consumer defined not by a fixed reality but by ongoing interaction.

Humanity's love affair with paper books stretches back into the mists of time, way before 20th-century culture imposed its values; but we cannot be sure that someone born at the beginning of the 21st century will have any particular nostalgia for a paper book. On the contrary, a book may be a novel object to someone who has had a screen-based education. Then again, the very nature of books will be different. We saw in Chapter 2 that within the next decade or so technology will have left its mark on books: screen resolution will be so improved that we will be able to read off the screen as easily as from paper books. However, preserving the romance and convenience of hand-held paper books, the digital age will also provide a download facility. E Link Systems and Xerox are developing thin films of paper and plastic that hold digital ink; once you finish a particular book you simply place it back in a holster for the next load of material. The same sheaf of pages would offer an eternal supply of different reading matter.

But perhaps you feel that to concentrate on a book as a physical object is to miss the main point: books are a very special phenomenon for a different reason altogether. Independent of their cultural connotations, they alone can foster and tap into our imaginations. As a neuroscientist I have long been fascinated by the process within the brain that transports us, who are simply staring at the written page, into a Victorian drawing-room, or a spaceship, or some fairy-tale scenario. So real is the world we imagine, primed by mere words, that we will almost always claim that the book is much better than the film of the same story. Indeed, when the film follows a good book we often feel a little cheated. Somehow, the characters are not quite right. Everything is too literal, too in-your-face, too reliant on sensory information. The fascinating feature of words, after all, is that when we read them they are shot through with connotations that transcend mere description; they stimulate the covert connections that give a deeper meaning. Hence, when we read, the characters do not necessarily have a photographic substance but one even more real, even though their physical appearance is shadowy.

This fascinating phenomenon of human imagination may prolong the popularity of works of fiction as books, as start-to-finish reads, that are in any event more convenient on planes or on the beach. Of course, such a prediction may hold good only for the immediate future. We cannot presume either that our successors thereafter will be capable of the same feats of imagination or indeed that computers will continue to be bulky and inconvenient, since we have seen that IT will become both ubiquitous and invisible. We have also seen that humans and computers will be more intimately related and interactive than we could ever have dreamed possible; it is most likely that the isolated, private inner world of the individual imagination as we know it could soon be as obsolete as the ability of our ancestors to recite tribal sagas from aural memory is today. Perhaps future generations will no longer have the attention span or cognitive skills to follow the narrative of a story. Perhaps, in the future, humanity will be rooted incessantly in the here and now.

Already, three-quarters of American high-school students prefer researching on the internet to using reference books: People of the Screen indeed. Moreover, there is some evidence that computer-assisted learning is beneficial. Mid-school students in rural Georgia who have been given laptops are showing improvements in grades and attendance rates, whilst many of their siblings who had dropped out of school have returned. One reason for this trend could be that learning is optimal when interacting – computer learning, using keyboard, screen and mouse, enables more interaction than learning in the standard classroom.

On the other hand, the arrival of computers in the classroom has not been accompanied by any statistically significant improvement in pupils' academic achievement. Some warn that we should learn from the mistake of the failed IT investment in the workplace in the 1980s. A key issue that was overlooked some twenty or so years ago was the attendant need to spend two to four times more on training than on the technology itself. In fact, three to five times greater investment apparently needs to be spent on organizational restructuring and job redesign than on the actual equipment. But finding funds is already a huge problem for educational establishments, not to mention the difficulties of training every teacher.

Schools in the Information Age will not simply be the same kind of schools that we went to with merely more computers in them! One of the most fundamental changes in the future, as we saw earlier, will be in oral communication with, and personalization of, machines. Children will grow up interacting with dozens of 'personologies' emitted from their PCs, and therefore will be as comfortable communicating and socializing in the cyber-world as in the real one, perhaps even more so. The brain is particularly plastic, impressionable, when it is developing. Early exposure to computers, hypertexting, mouse manipulation, menus and binary decisions will inevitably leave its mark on the nascent synapses. A recent finding widely reported in the news was that young people have already developed thumbs as dextrous as their fingers, due to incessant playing of the Gameboy and text messaging; believe it or not, children are even starting to point with their thumbs! So surely any day now exams, like all school activities, will be converted to a screen-based medium. This change will be welcomed even by the current generation, increasingly used as they are to assimilating information using a keyboard and screen, forced to sit for three hours with a pen and paper and expected to express themselves by increasingly obsolete and arduous handwriting.

However, we have no idea whether this new type of environment will be ultimately beneficial or deleterious. It could be the case that multimedia stimulation, assaulting the senses, hard-wires the brain for faster cognitive processing. On the other hand, what about reflection and imagination? Will the ready-made, second-hand images obviate the need or opportunity to think up one's own? Will the urgency of the multimedia moment keep the next generations ruthlessly reacting to the present, with no time left to retreat into daydreams?

But perhaps we should not judge new minds by old values. Since the essence of the human brain has been, for tens of thousands of years, adaptability to new external demands, perhaps we should simply face the fact that the new generation of brains will be fundamentally different from ours, in that they will be specifically suited, cognitively and physically, to computers and a cyber-world. Of course, for those born at the beginning of this century technology will be changing even faster; more than ever, succeeding generations will need to adapt to technical innovation. However, one fundamental question – given the potential of the screen to protect upcoming generations from the hectic, random heave of humanity 'out there' – is whether young people will be able to integrate material that they can understand intellectually but not necessarily appreciate emotionally. Will the new way of life in the 21st century mean that young people are more mature, or less?

My mother, born in 1927, often used to tell me that in her day there was 'no such thing as teenagers'. In fact, the concept is still unknown in some less developed countries, and it was indeed only in the post-war world of the second half of the 20th century that teenage culture started to flourish. Before then adolescents used to be apprentices, students, soldiers and farmers, but not teenagers. Reform of the child labour laws in the 1930s, the spread of suburbia and targeted youth marketing in the 1950s combined to give rise to the culture, or cultures, of the iconic dress, music, speech, ideas and behaviour that each of us remembers as 'special' in our own era, and, of course, superior to anything else before or since.

Yet it may be that adolescents are doomed to return to their original obscurity as the teenage age gap is closing. Certainly few would dispute the precocity of the young of today, let alone of the future. The average age for starting menstruation is now twelve, compared to fifteen in the 1800s. There has also been a dramatic increase in teenage mothers – the UK currently has the highest number of unintended teenage pregnancies in Europe, and in a recent survey almost 38 per cent of girls aged fifteen admitted to having had sexual intercourse. A conspicuous consumer culture geared, via magazines, adverts and music, to steering ever-younger girls towards clothes and boyfriends has been blamed by many critics for causing this change. And this is not so much 'growing up', in terms of having more control and insight over your feelings, as simply losing one's childhood at an earlier age.

Yet the demise of teenagers is not entirely due to the increased potential for IT-agility and independence at a pre-teen age, but also rests on the notion that the process of getting older is slowing down in grown-ups. Certainly, the introduction of lifelong learning, and its necessity in the workplace, means that adults need to preserve their learning skills as much as possible whilst somehow still using their previous experience to best advantage. These contradictory demands of being open to innovation but at the same time being able to evaluate new people, processes or things in the light of experience could well cause emotional turmoil and distress. But the central issue here is that adults will be behaving, or trying to behave, more like young people for longer. Undoubtedly, the advances in healthcare and health awareness, combined with an increased flexibility in choosing a life partner, will blur the distinction between the teenager and the twenty- or even the thirty-somethings, just as advances in healthcare will blur the distinction between middle and old age.

But irrespective of blurring age boundaries the problems tradition-ally faced by teenagers, and now by those behaving like teenagers, will not go away: concerns of identity and introspection will, if anything, probably increase. According to the 1995 Youth Risk Behavior Survey, 30 per cent of American high-school girls reported seriously consider-ing suicide, along with 18 per cent of boys – indeed it is the leading cause of death in this age group. There are increasing expectations to study and perform better, plus peer pressure to have sex, with the attendant risks of HIV and early pregnancy. Making your own decisions and sorting out your own system of values has never been quite so important as for the young person in the modern cyber-shadowed family, nor as difficult – given the diminution in parental influence coupled with the decline, and perhaps eventual demise, of the notion of a constant and unambiguous individual self – especially when it comes to that traditional teenage obsession: relationships, more specifically their initiation – flirting.

As with its simple forerunner, text messaging, the appeal of cyber-flirting seems obvious. It makes you feel connected, part of an atavistic type of tribe, even, of other young people. There is also a feeling of immediacy and hence excitement, as well as the novelty of the un-known. A more sinister and deeper concern, however, is an issue that has been cropping up time and again as we ponder what the future holds – the issue of identity. One cyber-flirter claimed that they enjoyed the activity because it 'makes me feel like a different person'. The staccato texting style can conceal a multitude of difficulties with inter-personal skills: concealment behind a persona of sophisticated software would be the natural next step for the fledgling adult nurtured by the cyber-world. And as with the prospect of a fictional cyber-family, so the extension of cyber-flirting could be flirting with fictional lovers. As a sign of the times, and also as evidence that such predictions are not too outlandish, I was amazed to discover that such a service already exists: for only $2.40 a month a Japanese mobile-phone com-pany arranges messages from virtual boyfriends – computer-generated, completely fictitious characters.

Just imagine how this trend may develop a few decades on, when teenagers, currently the most IT-agile generation, with the most leisure and freedom from professional and personal commitments, no longer have a monopoly in any of these areas. We might be faced with a society without the 'real' courtship that has characterized Western society for centuries, with all the traditional angst, fun, plotting and suspense replaced by a more anodyne activity where the individual is less vulnerable but, by the same token, not as fulfilled.

'Children's physical space is getting smaller at the same time as the cyberworld is getting bigger,' warns Judith Wagner, a child develop-ment specialist from Whittier College, California. Young people are spending more time indoors in front of their PCs; 'nature' is now something they see on a video. Then again, more interaction with the physical world may be possible, though from a remote and safe distance. For example, in 1994 Ken Goldberg developed the Mercury Project, which combined the web and robots to enable a network of individuals to operate, via their PCs, a robotic excavator. This robotic arm was guided, remotely, in an attempt to locate real items of 'treasure' buried in a real tub. He followed this with the more team-like sophisticated activity of 'Telegarden'. This time a real, six-foot-wide tub was divided up between a network of PC-users who each cultivated a small patch via a robot arm. This is the first true manipulation of the outside world from the cyber-world, inside, viewing reality indirectly. Similarly, though on a sweeping scale, the web has already made it possible for anyone to observe planet Earth live on screen, as shots are relayed in real time via satellite. 'I felt like I was seeing God,' enthuses Technophile Mark Pesce.

The impact of the Information Age is not simply that of the psychological revolution and evolution of a cyber-friend but rather the raw reality that your computer will connect you intimately with the whole planet. Some have even developed the somewhat fanciful metaphor of the Earth as a brain, with humans as previously isolated cells and the connections between them now newly forged as the net. In any event, such cyber-globalization means that we are due for some big cultural changes. Along with the facility for translating languages in real-time and access to worldwide news, there may even be a 'cybrarian', which will use voice-recognition to find educational materials that students need on the internet.

Moreover, information will be organized in the non-linear, more hypertext style of free association. And such associations will be expressed in visual media to convey experiences, in stark contrast to words, which convey ideas. So the next generation may well have more visual sensibilities, and be as proficient at manipulating images as their parents and grandparents – us – were once agile with words. Once literacy is truly as outdated as the slide rule and log tables are today, education will be transformed entirely into an *experience* rather than a thought process. For example, the mandarins at the BBC already foresee screen stuntmen using Pythagoras' theorem to calculate a trajectory for falling, a fire-eater demonstrating the nature of the elements, and 3D animation of atoms. But graphic and lively though such material might be, compared to the ancient talk and chalk methodology, there are drawbacks.

The first problem is that emphasis on interactivity would blur the distinction between the new tendency for a visceral, immediate response and a thought-out opinion. If a student expresses a view, how will it stack up and register against that of an expert? How will such a comparison be made? In the long run there may not be any 'experts' as the notion of a corpus of knowledge will have become obsolete. As the children of the future no longer need a long attention span to follow a linear narrative of words but rather are trapped in the immediacy of the 'now' – ever-stronger flashing lights and bleeps may be needed to sustain motivation or concentration over time frames of seconds.

Secondly, this noisy, bright and fast-moving display on the screen, transformed at the touch of a button, or the screen, at any moment you choose cannot help students work out abstract concepts, just from instant, screen-based evidence flashed up as a literal image before their eyes. Indeed, there is a risk that the use of new technologies for education will shift the focus to passive 'fun', indistinguishable from the rest of our sensorily overloaded cyber-lives.

But then again, we saw at the outset that interaction has huge potential to be educative. Take, for example, Lego Mindstorms, computer-controlled, programmable Lego bricks. Apparently, when Mindstorms was undergoing trials the children used for the testing couldn't believe that what they were doing might soon be regarded as school-work: it was too much fun. The Mindstorms bricks needn't be attached to a computer. Rather, the software is downloaded to a brick that then runs the programme. Mindstorms bricks seem to have struck just the right balance: there is enough pre-packaged technology without hampering creativity. The child, and indeed the interested adult, learns by doing. In general, interaction seems not sufficient in itself but rather a necessary starting point from which to develop the best methods for teaching.

The advent of nanotechnology, which promises unprecedented control over matter, could also inspire the logical successor of

Mindstorms. Mark Pesce predicts that nanotechnology could inspire a similar 'toy' for manipulating atoms. Children are already learning with a computer coupled to bricks; so the hypothetical child of the future might mess around with a plastic nut – essentially a pile of atoms – linked to a computer. It would be relatively straightforward for the child to create a ring of carbon – benzene – then to apply basic mechanics at this atomic level to end up with, say, a molecular calculator. The big question is, of course, whether nanotechnology will ever deliver its potential and achieve this type of precision; many are understandably cynical.

Nonetheless, a scaled-up version, at least, might be feasible, whereby components are assembled not on a nano- but on a micro-scale, instructed by the appropriate software. The Playstation 2, launched in November 2000, has proved the fastest-selling consumer electronics product in history – it is offering a glimmer of just how versatile cyber-toys will be in the future. As well as playing games the magic box can play DVDs and provide PC-type connectors. There are film-like simulations of the human face, as well as an image scanner and an interface with a camcorder. The user can thereby insert images of themselves as game characters. We saw that future readers – perhaps they should now be called simply consumers – may be able to determine the course of a novel and make an active contribution to the narrative; so here they are starring in a game. Once more the line is blurred between the individual who has created a work and the person who is reading or watching – the barriers between one mind and another are breaking down into a kind of all-encompassing networked brain. Moreover, as children and adults alike participate in novels and games, and as those same users have less and less practice at abstract thought, less imagination and less time for reflection, so there is a risk that the significance of facts and the desire to understand what is happening to and around you may diminish.

So perhaps education of the future should emphasize context instead of facts, since these can be readily accessed rather than learnt by rote; homework may consist of placing those facts into different conceptual frameworks – hypertexting will come into its own. In essence, learning as we know it may vanish in favour of a free-association hypertexting that is gradually rationalized and expanded. So, a linear knowledge, say, of the Tudor monarchs of England, with related insights into the literature and history of the 16th century, will be replaced by the term 'Henry VIII' cross-referenced to obesity, syphilis, divorce, gender selection, as well as to marine warfare, Martin Luther, Hampton Court, red hair and other less obvious associations. But a child will no longer actually 'know all about' Tudor England, nor 'understand' the factors that drove the Reformation in Europe, nor even perhaps have a grasp of the general concept of, say, religion.

Moreover, in order to hypertext at the moment we all need to have an appropriate knowledge-base first. How will that knowledge be obtained in the future? Information is not the same as knowledge, and somehow core concepts will have to be in place in the young mind in order for them to ask the appropriate questions regarding incoming information. Our children and grandchildren may well be able to roam the planet and interact with it from the other side of a screen, and catalogue facts pertaining to acid rain or depletion of the ozone layer with far more authority and cross-referencing than we ever could. But they will perhaps never take time to reflect on ways of putting those facts together in a way that we would currently characterize as understanding, at the very least, and as a creative idea, at the very best.

A further concern is whether children of the future will ever need or want to move from the wide, sanitized world opened at a touch of the keyboard to a more visceral experience within the confined limits of a real back garden, street or park. As education becomes an ongoing experience, and therefore less differentiated from everyday life, and as that experience is increasingly screen-derived, perhaps not just the notion of 'learning' but even the traditional concepts of 'school' and 'university' will start to become meaningless.

If the maturing of the Information Age is to revolutionize all aspects of education from *how* we learn and *what* we learn to what we learn *with* then it is no surprise that another change might be *where* we learn. Glen Russell, Lecturer in Education at Monash University, sees three types of virtual schools on the horizon. First, the 'Independent', where students access and interact with materials whenever they wish; such a school would not rely on real-time communication between student and teacher. Second, 'Synchronous' schools, where scheduled online meetings take place with other students and teachers, consisting of live chats and video-conferencing. This system would offer more socialization but reduced flexibility in timing. Thirdly, 'Broadcast' schools, where students would access lectures or broadcasts on the web; the biggest disadvantage here, of course, is that interaction would be restricted.

In the future, needless to say, all three types of school could combine. But there could, in any event, be a big problem when students work from home, using the internet, without ever being part of a conventional classroom. It may well be that those working remotely won't be just those who are physically unable, through disability or geographical location, to attend conventional school; an increasing number of 'regular' students will be attending virtual schools simply because they wish to. There is the very likely prospect of a dual education system. But the big issue that virtual education throws into focus is what a school is actually for.

The development of real face-to-face human relations, many would argue, is as important as learning facts. The consequences of switching to virtual schooling could be that students will fail to develop an understanding of their own emotions, and those of others; the patterns of lifelong friendship will be reduced; and mere facts could take precedence over the wisdom that comes from real experiences and spontaneous dialogues. The prospect of a solitary student working alone on a computer for their entire education is not intuitively appealing, but perhaps virtual schools are inevitable nonetheless. Our current education system could be viewed as a product of the Industrial not the Information Age. Students are currently subdivided into classrooms and year groups, given standardized textbooks, made to memorize information and regurgitate it in tests. This grading, like factory quality control, fitted the needs of the 19th century very well. In the UK, at least, the traditional public school, with its anti-intellectual culture, its emphasis on team games and group leadership, along with physical discomfort and abysmal food, equipped its pupils perfectly to run the remote and inhospitable corners of the British Empire – to work closely with others within a rigid order and hierarchy that was never challenged.

But now . . . Apparently the sum total of knowledge in the world –

or perhaps I should say information – is doubling every four years. Multimedia are being developed with text, sound, photos and video that could tailor the material on a syllabus to an individual child's learning style, be it association, abstract, visual or whatever. Moreover, the buildings of traditional schools are costly: one estimate for the USA is that it needs to invest $112 billion to repair or upgrade 80,000 school premises. So perhaps there is a persuasive argument for dispensing with one-size-fits-all curricula and, indeed, one-size-fits-all bricks-and-mortar schools.

Different models are possible for a virtual school of the future: perhaps a dense neighbourhood of sub-network schools, or block schools with ten children or fewer, or simply home-schools where a child goes at their own pace. Approximately 150,000 children in Britain are home-schooled, and this figure is predicted to triple by 2010. This hyper-localized learning, perhaps with parents working from home themselves, or with friends or neighbours wanting to teach in the context of a particular religion or culture, would necessitate socialization through different methods, hobbies or sports. The obvious criticism, that the minds of the young will be narrowed within their culture, or even within their families, could be offset by the added advantage of distance learning with children from different cultures all over the world.

Since specific facts will be accessed on a just-in-time basis, emphasis will ideally be on *how* to learn – mastering concepts, thinking critically and expressing oneself effectively: no less than a preparation for the lifelong, independent learning that the lifestyle of the second half of the 21st century will entail. Following this theme, perhaps formal education will end much earlier. Learning will be less associated with institutions, and more with need-driven learning in the workplace.

Future generations will not therefore be generalists. The Information Age could lead to more specialization not less, as the specific needs of diverse aspects of society are met by appropriate technologies, ever-changing and needing ever-changing skills, as we saw in Chapter 4. We could be on the brink of returning to the mood of the mid 19th century, when most of the workforce were apprenticed at a young age and learnt one 'vocational' skill that fitted the needs of society. The only difference compared to the old system would be that within a certain trade, or vocation, constant retraining – lifelong learning – would be an accepted feature of everyday life. And the skills would not be manual but invariably IT-related.

The problem with an IT-based education, with the focus on the individual going at their own pace for their own individual needs and curiosities, is that surely there will be an inevitable loss of direction in terms of what we are learning as a cohesive society. If, rather than thinking, everyone is having an experience with interactive and personalized pedagogic programmes, how might we make progress in increasing our common, shared knowledge-base and our progressive quest for the truth? The cynical might even say that this new type of cyber-education will hold out the lure of a quick fix for problems of pupil attainment. Everyone will be a winner; an individual curriculum might be the ultimate dumbing-down.

Distance learning has the undoubted advantages of offering frequent opportunities to experiment with new technologies, minimal instructor travel, access to guest speakers, along with just-in-time training that couldn't be achieved previously, since conventional classes required preparation done a week or so in advance. A further advantage of multiple-conferencing technologies is that they might actually be preferable to face-to-face communication, since one would get to know more fellow students than just those sitting nearby. The fact that students seem better able to fit classes into their schedule, plus the fact that there are shorter classes scheduled over a longer time, may account for why the drop-out rate from such classes so far has proved very low.

But what about this new technology applied to universities? Already almost two out of three American children go on to higher education, and this population is set to swell to 16.3 million by 2009, according to the US Department of Education's figures. Currently, the over-25s account for some 40 per cent of the student community. In a few years, mature students over 35 will exceed 18- to 19-year-olds. Virginia Tech already offers a 24-hour-a-day programme in undergraduate maths taught electronically. Students work in a communal space with computers whilst faculty and teaching assistants roam the aisles rather than leading the class. A 32-year-old student is less than enthusiastic: 'This is Orwellian Math. It's just you and the machine, and the professor is this shadowy figure who emails you once a week.'

But then, test scores and enrolment are up, with fewer drop-outs, and the costs are down. Like secondary schools, there could be various types of virtual universities. First, public educational offerings such as those developed already by the Universities of Texas and California; second, collaborative educational efforts between universities such as Western Governors virtual university, a collaboration of most western states excluding Texas and California; third, private entities, such as the University of Phoenix with 60,000 students; finally there will be an increasing number of corporate training establishments, as exemplified by the Marriott corporation, which offers world-wide training programmes as well as educational services to other businesses.

Once again, even more perhaps than at secondary school, we have to ask: if students can work at their own pace and direction to an extreme degree, what then constitutes a course? The notion of personalized degrees is strange, but might be a feature of the future academic landscape. Given the individual needs, interests and abilities that can now be catered for, and given that computers could structure and mark tests, there is no technical reason why we all should not take this path of ultra-individualization. There would have to be, of course, some way of assigning a value to a particular piece of knowledge or experience. In addition corporate degrees might become popular, with exactly the right blend of skills for each company, or standardization across companies who produce their own modules for assessment.

Universities in the future therefore will be not so much places where you gain knowledge for life as engines for technological advancement. Yet Jim Dator of the University of Hawaii sees the change as inevitable, and bricks-and-mortar universities as relics of the past. He argues that although universities are very old, the public education system has been going for only 150 years and was designed to meet the needs of emerging industrial states. Formal education was not needed in the slow-moving, agricultural, feudal economies. But then, farmers and peasants were transformed into workers and managers. The idea, from the point of view of the state, was never to enable scholars to 'pursue truth'.

As we enter the 21st century, in education, as in life generally, space and time will become less and less standardized. Students will be learning at different rates in different places. The international dimension of new technologies will enable a sharing of different cultural perspectives. Western thought will no longer dominate, but is predicted to come fourth behind Confucian, Hindi and Islamic agendas. In addition, students will gain experience from simulated real-time experiences in virtual hospitals and factories. But research might be cut, unless there is a direct military or commercial spin-off. Meanwhile campuses may well become theme parks: Bill Gates has already funded such a venture at Harvard, where the scene is set permanently in 1925, complete with lectures on the topics of the period, such as Marxism and relativity theory. Imagine, then, the future education of your grandchildren or great-grandchildren, inevitably in some type of virtual university, on a course that is highly tailored either to them personally or to the needs of their employer. David Waguespack of the University of Oregon sums up the situation:

The virtual university is truly a mixed bag. Among the benefits are competition, choice, and greater educational access. At best, the virtual university will force regional colleges to improve and rethink teaching, because alternatives will be available. Among the downsides are turning education into a commodity and a degraded college experience. At worst, the virtual university will create a situation where credentials take precedence over learning, and educational convenience masquerades as greater access. The truth is likely somewhere between the two extremes. What ultimately comes of the virtual university depends not on the opinions of academics, however, but on whether the consumers of education want it.

So what do we want of education? The most gloomy prediction is that we will be living in a society geared to the material needs and desires of society, albeit a global one, where time and space have little relevance. We will inhabit a world of experience, more specifically, screen experience, rather than abstract thought; answers will crowd onto our screens and compete for attention, no longer linked to any clear questions. There may well be nothing about our new world that we need to ponder. Universities will no longer be a central plank in our culture, primarily because no one will believe any longer that 'the truth' is out there waiting to be discovered, let alone that it is beautiful. Is this intellectual heresy really what awaits us?

7

Science: What questions will we ask?

I think people get the wrong impression about scientists in that they think in an orderly, rigid way from step 1 to step 2 to step 3. What really happens is that often you make some imaginative leap which at the time may seem nonsensical. When you capture the field at those stages it looks like poetry in which you are imagining without yet proving.

Paul Steinhardt, physicist.

Having looked at how we will live, work, love and learn, the time has come to ask what the future holds for the human imagination: how will our successors cope when, and if, it comes to tackling new, big questions of science? So far our entire journey into the forthcoming decades of the 21st century has been driven by the startling advances in a vast range of technologies, but these dazzlingly innovative incursions into our lives actually have their inception in basic concepts introduced in the 20th century – computers and genetics.

As it happens, computers and genetics, though seemingly unrelated, do share a common origin, a single leap of imagination in basic science almost a century ago. That colossal intellectual milestone was quantum theory, pioneered by Werner Heisenberg and Erwin Schrödinger in the 1920s. Quantum theory challenged the idea, seemingly impregnable at the time, that waves and particles were distinct and suggested instead that they were inseparable. Heisenberg and Schrödinger used the notion that waves and particles were really two sides of the same coin to describe the 'quantization' of energy, the process whereby energy can be transferred only in packets and not, as had been thought until then, in a continuous manner. Abstract and baffling as quantum theory may sound the insights it gave into the basics of matter and energy were to have astounding implications for more down-to-earth branches of science. Advanced materials such as lasers and transistors, and therefore ultimately computers, rely on the principles of quantum theory. Likewise in biology, the currently emerging feats of gene manipulation, triggered by our ability to manipulate atoms, are reliant on an understanding of molecular bonds and the technique of X-ray crystallography, both of which hark back to quantum theory.

These knock-on effects of quantum theory have provided enough work for generations of scientists and technologists over the previous century, and keep their contemporary counterparts fully occupied well into this one, contemplating its further resounding implications. But was quantum theory, and the various scientific revolutions it spawned, a one-off? Some think that we will see no comparable great breakthroughs. The science journalist John Horgan, author of *The End of Science*, for example, claims that: 'Scientists will continue making incremental advances, but they will never achieve their most ambitious goals, such as understanding the origin of the universe, of life and of human consciousness.'

How valid is this view? When we looked at future education in the previous chapter the threat loomed large of a new way of life, one that emphasized the passive, the hedonistic and the experiential over abstract thought and imagination. Such a state of mind, obviously, would doom any further original scientific endeavour. But, for the sake of argument, let's assume that somehow human imagination manages to survive the suffocation of an utterly comfortable lifestyle. If indeed highly innovative ideas and insights that challenge existing thinking always take time to be translated into practical technology that affects everyone's daily existence, then the thoughts of scientists of this century will be setting the pace and agenda for life well into 2200 and beyond.

Let's think about who those scientists will be. The rugged individualists of the 19th century, such as Michael Faraday and John Dalton, who set up labs in their cellars, scrabbled around rocks or simply observed, like Darwin, the world around them, eventually gave way to the more institutionalized genre of academic scientists, funded by governments and charitable trusts to give, as Haldane predicted long ago, 'the answer of the few to the demands of the many for wealth, comfort and victory'. But nowadays the horizons of the Ivory-Tower boffin are being widened by two new trends. First, there is a growing need for innovative science in the private sector, as companies in high-tech industries, particularly pharmaceutical companies, depend for survival on having novel products in the pipeline. Allowing for a rosy enough economic situation – a big assumption – there will therefore be a need for more ideas-based companies; this drive for intellectual innovation could result in a smorgasbord of new opportunities for the development of intellectual property within universities. If entrepreneurs regain the confidence and capability to invest in high technology, then science will prove a more lucrative, exciting and, above all, useful career, and it may become a profession as structured, and as popular, as law or medicine within the next few decades.

The second recent trend is the resurgence of the amateur. As an example, take the Royal Institution in London, where I am Director: it was founded in 1799, in the words of its charter, to 'diffuse science for the common purposes of life' and the idea was to ensure that lectures on the scientific discoveries of the time were as rewarding as any other evening outing. So successful were these presentations by the great scientists of the day, complete with bangs, smoke and other exciting demonstrations, that the lecture theatre in central London rapidly became a fashionable salon, a place to be seen as much as to think and to question how science – for example, in the shape of the newly invented electric motor – was impacting on and changing the accepted way of life. We are now seeing a revival of this attitude. For the last few decades popular-science books and science broadcasts have captured the public imagination. Someone has even remarked that science is now like foreign travel was in the 19th century: because it was not accessible at first hand people were keen to pursue it indirectly, through the eyes of an intrepid but always articulate pioneer. Science debates, and events such as authors of science books lecturing at literary festivals, attract hundreds if not thousands at a sitting; above all, the public are realizing that they need to be empowered with knowledge, to be scientifically literate, if they are to contribute to the great debates that science will inspire this century.

The combination of private-sector funding plus a more benevolent and informed science-friendly electorate could, if neither cyber-hedonism nor politically related science scandals prevail, create a chance for scientists truly to exercise their imagination. Indeed, scientists can now crack on as never before: automation has freed them from routine lab drudgery. All manner of databases are now only a key tap away, whilst whole new branches of science are opening up *in silico* – in the silicon memory banks of computers as opposed to in a lab dish (*in vitro*) or most vanishing of all *in vivo* (in a whole animal). In the future instead of performing a real experiment you could 'mine' data – tease out a line of enquiry from the colossal monolith of facts 'out there'. Or you might choose to experiment, but on a computer model whereby you can test out a drug on a cyber-heart, or observe other types of biochemical and physiological reactions that, in the old days, would have necessitated setting up a messy and capricious real-life trial.

Like all other aspects of our lives, then, the Information Age has changed the world of scientific research – it is easier to work within, gives answers quickly and reliably, does not demand great physical dexterity or lab-savvy green fingers; scientists now work in a world that saves so much time that even the most daunting information-crunching tasks suddenly become possible. Meanwhile, the fuelling of technology with the latest scientific findings will be more and more voracious – delivering the objects of desire, all the gifts of 'wealth and comfort', that have formed the material of the preceding chapters. In the universities as well as behind the walls of the leviathan pharmaceutical and other high-tech industries automation will take away the ingenuity, the luck, the wrestling with a capricious physical world that has characterized research so far, in favour of a frog-spawn approach – all strategies, be they more or less likely, can be tried out regardless by machines. The lack of any insight on behalf of the robot researchers is more than offset by their sheer speed at churning exhaustively through all possibilities, coming up with the right answer in a fraction of the time that it would have taken their more reflective but fallible human counterparts.

Now let's think about *what* exactly a full-time scientist, freed up from more routine occupations, will pursue as a big new problem. The theoretical physicist Freeman Dyson has suggested that innovative science can be classified into two categories: first, to explain old things in new ways, 'concept-driven revolution', such as the ideas of Einstein, Darwin or Freud; secondly, to discover new things that have to be explained, 'tool-driven revolution', for example, Galileo's use of the telescope, or the exploitation of crystallography in molecular biology. How far will the 21st-century scientist progress with either type of revolution?

As a prelude, it is most likely that the traditional disciplines of science will not be recognizable as such, but will have merged. For example, computer science as a subject originally arose from diverse areas including mathematics, engineering and physics, whilst the modern study of the brain, neuroscience, is a product of many different disciplines including computer science, physiology, anatomy, bio-chemistry and psychology. Yet neuroscience itself may well end up lumped with electrical engineering and molecular biology to give rise to a new discipline, 'neurotechnology', in which future generations of scientists do not just study the brain, but probe, display and manipulate it with an immediacy and precision that will have far-flung ramifications for how we live.

Meanwhile computer power will soon be such that there could be an exhaustive database relating to all aspects of brain operations, to the whole multidisciplinary subject of neuroscience, that had previously been segregated according to a particular expertise in, say, pharmacology or physiology or genetics. With this great store of multidisciplinary information available even a non-specialist will be able to rove across all subjects relating to the brain, pursuing a pet theory and having it automatically cross-referenced, validated against all known experiments and used to make further predictions. Just as the geneticists are already engaged in this type of silicon exploration with bio-informatics, so we could develop an analogous process of neuroinformatics. In short, 21st-century science, its biology at least, will probably be characterized by a swing away from the reductionist tendency, the analysis that was the hallmark of much of research in the 20th century, in favour of synthesis, describing inter-relations between previously disparate areas, and indeed of developing umbrella concepts that transcend those particular disciplines.

Cross-referencing in this way could give new insights and inspire the next 'concept-driven revolution'. Some of the greatest science has arisen from interactions at the frontier of classical cultures. For example, Linus Pauling, in the first half of the 20th century, imported the principles of quantum mechanics into chemistry and revolutionized our understanding of the chemical bond. Another cataclysmic insight came to Sir Frank Macfarlane Burnet in 1959 when he realized that the principles of Darwinian evolution could be applied to a completely different domain, the immune system. In the future, such leaps across disciplines may become increasingly frequent as we have more time to survey, from the hilltop vista of the internet and the world wide web, more subjects; with increasing ease we will be able to harness IT to seek similarities in patterns and deep organization, independent of the scale or speed or jargon of any one particular subject.

So, the merging and cross-referencing of different branches of science will provide a fascinating opportunity to identify fundamental concepts, analogous to, say, the survival of the fittest. But such ideas can come only if scientists ask the right kind of questions first. In the future, then, what will need to be explained or discovered? In 1923 the biologist J. B. S. Haldane delivered a paper on the future of science to a Cambridge society, The Heretics. He called his vision – which was to inspire Aldous Huxley's *Brave New World* – *Daedalus*, after the father of Icarus in Greek mythology, who did *not* fly too near the sun with waxen wings. Reeling still from the horrors of mechanized warfare, Haldane explored the future of science, which he defined as 'the free activity of man's divine faculties of reason and imagination'.

Many of the predictions in *Daedelus* turned out to be chillingly accurate, and actually articulate fears not unlike those we have been discussing here, nearly a century later. For Haldane, technology did not necessarily bring in its train some easy nirvana, and even if it did, then that in itself would pose a problem – indeed the type of difficulties arising from the unbridled comfort that we have been discussing. Just as we may have to square up to a new cyber-passivity so Haldane entertained the bleak prospect that humans might end up as a 'mere parasite of machinery', and predicted that physics would abolish not just the constraints of day and night but also other spatiotemporal checks to our lives – a vision now realized by the web and the internet.

Although Haldane envisaged that chemistry would continue to alter life as it had done before it was even a science, with explosives, dyes and drugs, it was in the application of biology that he foresaw truly big transformations. Looking hard at the nascent eugenics movement of the time, one near-certainty was a 'eugenics official' and 'marriage by numbers'. And although such phenomena have only come to pass indirectly, with virtual dating and the possibility of designer children, other predictions were startlingly accurate. As well as foreseeing the abolition of many infectious diseases, Haldane predicted the development of a 'nitrogen-fixing' plant that in a sense anticipated GM food. He even described a possible eco-accident, of the type all too familiar nowadays: the sea turned purple due to contamination by that plant. Haldane could also have written, though from the viewpoint of the 1920s, a credible version of my Chapter 5, concerning future views of life. He actually prophesied the development of IVF and the complete dissociation of sex and reproduction, with his concept of 'ectogenesis'. Even Prozac was on his horizon: '. . . to control our passions by some more direct method than fasting or flagellation.' And he scored a direct hit in anticipating HRT: 'This change seems to be due to a sudden failure of a definite chemical substance produced by the ovary. When we can isolate and synthesize this body we shall be able to prolong a woman's youth, and allow her to age as gradually as the average man.'

Haldane not only had a wide-ranging grasp of the science of his day but also an impressive instinct for what was to follow. I suggest we therefore turn to him to give us a framework for the most important issue of all in our current discussion. What of *fundamental* science, the science that will spawn the technologies of the next century: what are the remaining big questions here?

Haldane listed them in *Daedalus* as problems 'first of space and time' – in our terminology the big bang; 'then of matter as such' – for us the persistent quirkiness of quantum theory and the dream of nanoscience; 'then of [man's] own body and those of other living beings' – surely the synthesis of different branches of biomedical science, along with the greatest question of how a brain can generate the subjective experience of consciousness; 'and finally the subjugation of the dark and evil elements in [mankind's] own soul' – at a stretch, the big question of how we shall use this knowledge to work out accountability, the degree of biological determinism, the ultimate riddle of free will in the Neurotechnological Age. These same questions also feature in John Horgan's idea of the 'most ambitious goals' of science.

Imagine, then, that you are a scientist researching one of these questions a few decades from now. You like using your imagination in this way because it is the only conduit left for such mental indulgence, and because when you are coming up with original ideas it gives you a feeling – unusual now in the mid 21st century – of personal fulfilment. It is the most effective way for you to buck the current trend and feel like a complete yet independent person. You spend most of your time talking to your screen; when necessary you are able to assemble 3D models readily enough. And types of equipment that were once stratospherically expensive, such as a particle accelerator, or merely costly, like special types of microscope for the manipulation of atoms, or even the older technology of humble magnification of some 100,000 times to observe the components inside the cells that make up biological tissue, are all now yours at a voice command.

Most of the time the simulations are breathtakingly faithful to the real thing. Highly sophisticated programs covering all contingencies for any experimental situation enable you to get 'virtual data' almost as fast as you can think up and programme in the virtual experiment. More and more of the general public are therefore not only reading science books but doing science, performing virtual experiments. In fact, nowadays even most of you professionals only occasionally access the remote, real apparatus to verify the accuracy of the cyber-studies if the result is particularly surprising, and yet as yet the virtual data is not significantly discrepant from its counterpart, garnered so much more painstakingly, in the real world.

However, science remains the one activity where you still have to be careful to distinguish reality from virtual reality. All your simulations are built on models of interactions, on models of the forces acting between particles, and a preconceived vision of what exists. You know that is no substitute for hands-on 'real' science, which constantly compares its hypotheses with reality. Here there is a barrier to progress, for that examination – and the likelihood of unexpected discovery in the real world – will become prohibitively impractical.

Some theories about elementary particles might never be tested exhaustively, for to do so would require such colossal accelerators that they would have to span the universe and consume the outputs of all economies everywhere.

One investigation where observation is essential and simulation a guide, which encompasses Haldane's first big question concerning space and time, is that of the origin of the universe. There is no doubt in the minds of cosmologists such as yourself that our universe came into existence about fifteen billion years ago and that it has been expanding ever since. Initially, the entire visible universe was confined to a single point, a singularity. But even before you reach that far back the current concepts of space and time break down, and in your current research you are struggling to extend scientific techniques into this region. To achieve understanding of the very, very early universe, a universe so small that quantum events dominate, quantum theory must somehow overcome its current failure to account for the force of gravitation that would have been operating: you need to describe a 'quantum gravity'. Like the rest of your colleagues, you really do not know how this will be done, although there are currently many suggestive ideas. Scientists, being optimists, believe that a theory of quantum gravity will be achieved so that we will be able to understand the earliest moments of the universe. But will science be able to go beyond that instant of creation and discover how the universe originated? You do not know. Some doubt that we ever will but others – you among them – are enthralled by the prospect of going back to before the beginning.

For example, you know that even cherished ideas that have been part of our civilization since classical times are likely to give way as science overturns the everyday in its quest for the truth. Up until the end of the 20th century everyone thought of the end point of matter as literally that: a collection of points. At the beginning of the 21st century an alternative view emerged, that the ultimate entities are tiny, rigid lines, known as 'strings'. A particle was seen as just a vibration of a string.

Are strings the end of the line, or can you expect the overthrow of even that concept before we reach ultimate reality? Most scientists think that there will be a final answer, a 'theory of *everything*', but no one really knows. These strings are thought to exist in ten, perhaps eleven, dimensions of spacetime. This bizarre–sounding concept was first introduced by the 19th-century mathematician Hermann Minkowski: 'Henceforth, space by itself, and time by itself, are doomed to fade away into mere shadows, and only a kind of union of the two will preserve an independent reality.'

Spacetime, as its name suggests, is a notion that puts space and time on an equal footing so that we are freed from parcelling out our lives, the world and indeed the universe in the obvious commonsense way. As our understanding of it has grown we have discovered new ways to travel through it, and new barriers to travel too. Time travel has always been with us: after all, we all drift inexorably into the future. However, Einstein's theory of special relativity showed that this inexorable drift is more complex than we thought, for the faster we travel through space the slower our clocks run. We could retard our ageing were it possible to travel close to the speed of light, but it seems that we cannot reverse time and grow young, perhaps discarding our memories as we do so. Yet the Holy Grail of quantum gravity might offer this possibility.

Neither you, a late-21st-century scientist, nor your colleagues can give a definite answer: but the ranks of scientists thinking seriously about using the foam-like structure of spacetime as a tunnel to travel faster into the future or even into the past are swelling. Ever since the beginning of the century scientists have been studying these tunnels, wormholes, and are viewing them less and less as science fiction. Travel through such tunnels is still fantasy for the moment, at least for anything bigger than an electron, but it remains a conceivable option for the future. The very notion of time travel is plagued by paradoxes; but now our notions of classical logic are being modified as regards the passage of time and the very nature of causality. Logic, though of extreme importance in science, cannot be regarded as superior to experiment and observation.

Nonsensical leaps of imagination, such as time travel, as well as more immediate, hard-headed questions to which there must be some kind of answer, about the big bang and the expansion of the universe for instance, all hinge on whether it will ever be possible to reconcile the two sets of physics – on the one hand the set of rules for subatomic particles, and on the other for the macro, tangible world. If you could discover some overarching framework, a 'theory of everything' for understanding the interaction of matter and energy with space and time, and when and how and what comes into play, then the extreme eventuality is that you would be able to unleash the seemingly bizarre quantum phenomena that defy time and space and causality into our macro everyday world. This would be the breakdown of the biggest barrier of all, wrenching at the fabric of reality. You are currently, in the mid 21st century, living in a world where nothing is of relevance any longer in the real world, because the cyber-world has insulated you from it. But what if that reality itself was now changed to one in which space and time, as such, no longer existed?

Let's leave the physicist trying to conceive of an independence of space and time, in effect a future version of Haldane's first big question. Instead, imagine now that you are a chemist, and as such more fascinated by Haldane's second question 'of matter'. Ever since the end of the 20th century your predecessors have been trying to manipulate atoms with an unprecedented degree of control. This new branch of science, nanoscience, continues to excite scientists and society alike, perhaps because it qualifies as both a 'concept-driven revolution', in that it opens up the possibility of seeing old things in new ways, but at the same time constitutes a 'tool-driven revolution', with the opportunity of making new discoveries.

The idea of nanoscience was first introduced in 1959 when the physicist Richard Feynman, later a Nobel-Prize-winner, delivered his famous lecture 'There's Plenty of Room at the Bottom'. Feynman argued that there was nothing in the laws of quantum physics that precluded the invention of machines the size of molecules. The basic idea was that eventually scientists could adopt a 'bottom up' approach, placing atoms exactly where they wanted them. The term 'nanotechnology' was coined a few years later by the Japanese scientist Norio Taniguchi, to describe machining in the range of 0.1 to 100 nanometres; hence nanoscience, if we adhere to the strictest definition, deals with materials or systems with at least one dimension less than 100 nanometres, i.e., less than a tenth of a thousandth of a thousandth of a metre.

However, in the early days some were sloppy in how they used this new term. There were those who used the word as a label for any device that was unusually small and thus might have an unusual job, such as acting as a minuscule submarine to monitor, and even scour, the furring up of arteries. But often these developments were really operating on the microscale, at least 1,000 times larger than a nanometre. It was important to distinguish 'true' nanoscience from the development of merely very small machines. Such devices, micro-electromechanical systems (MEMS), were nonetheless awesome – miniature sensors and motors about the size of a dust particle. These sensors and motors were etched onto silicon wafers using the same technique as in the microchip industry. Applications in those early days included an air-bag motion detector the size of a whisker, which allowed a reduction in the cost of certain laboratory equipment from $20,000 right down to $10, microdevices that could thread blood vessels, and cheap, pressure-sensitive devices embedded in steel and other building materials to detect the kinds of stress occurring in an earthquake, or on the surface of aircraft wings to detect stress during flight. However, after 2020 MEMS were ousted by machines on the truly nanoscale.

To get an idea of this scale, of just how small the nano-range actually is, consider that a human hair is an awesome 10,000 nanometres thick. The nanoworld, as the science writer Gary Stix pointed out, is the 'weird borderland between the realm of individual atoms and molecules (where quantum mechanics rules) and the macroworld (where the bulk properties of materials emerge from the collective behavior of trillions of atoms, whether that material is a steel beam or the cream filling in an Oreo [an American biscuit])'. As such, nanotechnology defines the smallest natural structures; put succinctly, it is impossible to build anything smaller. The burgeoning interest in the nanoworld has stemmed from the idea that structures on this small scale may have superior electrical, chemical, mechanical or optical properties. Once conventional silicon electronics ceased to work around 2020, then the new nanotechnology clearly offered the most realistic and attractive alternative.

It was in the last decade of the last century that scientists had started to take nanoscience seriously, following a breakthrough when scientists at IBM arranged 35 xenon atoms on a nickel surface to spell out their logo. Funding started to soar, even outside of the USA, rising from $316 million in 1997 to $835 million only four years later. Advocates of nanotechnology waxed lyrical about its potential, and indeed its impact on everyday life, university research and commerce in the 21st century. In the mid 1990s it cost $1,000 for the Nano-phase Technologies Corporation to produce a single gram of nanopart-icles; within a decade a gram cost only a few cents and could be used in products as disparate as odour-eating foot powders and ships. Within a few years nanotechnologists were going on to offer affordable solutions to complex and novel problems, such as the replacement of palladium, a costly component used in many cars for catalytic conversion.

All this direct application of basic science to industry fuelled a push for public money. In 1999 Bill Clinton announced $422 million for a National Nanotechnology Initiative; his successor George W. Bush announced a further $487 million in 2001. Governments around the world are estimated to be investing billions in basic nanotechnology research and development. This kind of backing was hardly surprising since basic research into nanoscience was tool-driven, leading immedi-ately to practical applications in virtually all aspects of life. Soon drugs were developed to detect and kill cancer cells before they could do any extensive harm, whilst Boeing 747 jumbo jets were quickly being built at one-fiftieth of their previous weight as new types of materials became available. New nano-agents rapidly became routine additions to vents of hospitals for detecting disease, whilst others attached only to cholesterol, so that they could be selectively targeted in hardened arteries. The domestic scene was also undergoing a revolution as new fabrics became available that were cool in summer but warm when the weather turned colder, and a new generation of household paints was developed to repel dirt. But the real excitement was not that previously everyday objects and products were suddenly 'smart' but that they had acquired astonishing, utterly novel properties. Now that surfaces of materials were layered with atomic precision, everyone soon became used to phenomena that previously would have seemed utterly imposs-ible, such as frictionless bearings, scratch-proof spectacles and more powerful fibre-optic cables.

Admittedly it had not all been plain sailing. One of the original worries was that, at the time, there was no appropriately scaled-down wire: after all, any potential molecular machine was useless as long as the different components within it were unable to connect with each other. This difficulty was overcome with 'nanotubes'. Nanotubes, still very like their turn-of-the-century prototype, are so thin that you would need 50,000 side by side to cover the width of a human hair; they are made by heating carbon to a vapour, then condensing it in a vacuum or inert gas. Amazingly, the carbon atoms then famously arrange themselves into classic, football-like hexagons – buckyballs – arranged in a long cylinder that not only conducts electricity but is some 100 times stronger than steel and weighs one-sixth as much.

Already, within half a century of Feynman's original vision, scientists had the eyes and fingers to manipulate nature's building blocks. The 'eyes' were microscopes a million times more powerful than the human eye. The prototype Scanning Tunnelling Microscope (STM) dragged a tiny needle across the surface of an object. Electrons could then 'tunnel' across the electrical barrier between the needle and atoms, creating a current that measured the position of atoms. Going one further, to act as 'fingers', the needle could move individual molecules to create stable hexagonal rings at room temperature by applying a voltage through its super-sharp tip. An additional refinement soon followed in the Atomic Force Microscope (AFM), which also probes surfaces and produces topographical images of individual atoms. These nanoscale fingers and eyes soon started to provide new knowledge about how matter operates at the atomic and molecular level. As a result it became possible to put together molecules that have never been connected before, not just nanotubes but nanotweezers for grabbing and pulling molecules; for the first time, scientists started to manipulate the different components within a single human cell. Now artificial chromosomes could be introduced with ease. Indeed, by 2030 sophisticated machines for even more precise manipulation of individual atoms were commonplace. In the first few decades of the 21st century there was no limit to the nanotechnologist's dreams: 'If nature has figured out how to arrange the atoms in coal to make a diamond, then we should be able to do the same.' So one expert at the time predicted, envisaging that within a mere twenty-five years there would be a new passion for alchemy, but with a scientific basis, whereby the wonders of nanoscience would transmute the banal into the special, as well as generating life-saving materials such as bones, spinal cords and human hearts.

But then the real problems started. Although the STM and AFM could move individual particles they were really too slow for mass production. Moreover, whilst careful control of chemical reactions could assemble atoms and molecules into small nanostructures of 2–10 nanometres, the process failed to produce designed, interconnected patterns of the type needed for electronic devices such as microchips. Moreover, despite Feynman having pointed out that there is nothing in the laws of physics that precludes the construction of nanodevices, not one molecular machine on the true nanoscale has ever, even now, actually been made in any lab. The problem has been that as micromachines gave way to their nanoscale counterparts the goalposts moved: the compliant and permissive laws of physics concede to the more capricious and non-negotiable ones of chemistry.

For example, the formation of bonds between atoms may also affect those atoms in other ways, which we cannot necessarily predict or understand. Quantum mechanics will probably prevail over events in small electrical devices where a single electron can dominate current flow, but the smaller a device the more susceptible its physical properties are to alteration. No longer does the physics of bulk dominate but rather the chemical properties of surfaces become all-important. In a silicon beam 10 nm wide and 100 nm long, as many as 10 per cent of the atoms are on or near the surface. Even nanotubes in the real world have electrical properties that are very different once away from their original and rarefied ultra-high-vacuum habitats. Clearly such a high proportion of surface molecules will play a central role, but how?

In general, we still cannot determine how arbitrarily assembled neutrons, protons and electrons will finally behave en masse. The really great challenge for nanoscience is therefore to find a way of assembling such small components together in a purposeful, controlled fashion. Unlike conventional circuit design, you find yourself beginning with a haphazard jumble of as many as ten to twenty-four nanocomponents and wires, not all of which will work and from which you must somehow gradually shape a useful device. Or to rephrase the problem: you need to think of a means of linking the nano- and microworlds.

For quite a while now these deep conceptual difficulties have tempered the initial, unconditional enthusiasm for nanotechnology. The chemist David Jones, way back at the dawn of the nanotech era, captured the mood of reality check: 'How do these machines know where every atom is located? How do you program such machines to perform their miraculous feats? How do they navigate? Where does their power come from?'

Although these questions are still unanswered, a core dream of some nanoscientists is nonetheless that nanomachines will be able to control themselves and be independent of interference from on high, from the human macroworld. The true convert envisages self-constructing machines, 'bottom up', with nanoassemblers and nanorobots. Such self-replication would result from gears and wheels no larger than several atoms in diameter, which would give rise to atomic-size machines that could scavenge molecules from their environment to reproduce themselves. These molecular robots, some tenth of a micrometre in size, could then manipulate more individual atoms. Such nanomachines could operate like bacteria, or viruses, for good or ill: they could destroy infectious microbes, kill tumour cells, patrol the bloodstream, and devour hazardous wastes in the environment, thereby enhancing the production of cheap and plentiful food. They could go on to build other machines, from booster rockets to microchips and supercomputers the size of atoms . . .

But perhaps most fantastical of all is the idea that nanotechnology could reinstate the original patterns of atoms and thus ultimately of cells that have been damaged, and so in theory, at least, reverse their ageing. You are highly sceptical, as are many of your colleagues, but the concept is straightforward enough. Ultimately, the frozen brains of the newly dead could be restored to their original atomic configuration by undoing the damage usually caused in thawing as ice crystals break down the cell wall. From this concept a whole industry, cryonics, has been flourishing in the USA since as far back as the previous century. The plan is that you sign up to be frozen once you have died, in the hope that once this technology of precision atomic rearrangement has been perfected, plus that of identifying the 'genes to survive', you might end up immortal.

Already there are many customers; however, the concept of a gene 'for' survival, as 'for' any complex process or trait, might be more than a little naive, as we saw previously. And even if we could discover Methuselah genes that we then used to enhance the genomes of our progeny, you, the cryonics 'patient' from a bygone era, would have missed out; in any event, the genes would not, even at best, guarantee living for ever, merely a prolongation of life. A further dependence on gene technology of more immediate importance is that in many cryonics schemes the patient is frozen as 'head only', in the confident assumption that the technology to grow a new body will also exist in the future. Let us hope that the nanotechnological reversal of damaged cells and revival from freezing do not precede that science of growing a new body. The prospect of disembodied but conscious heads suddenly joining the world population is surely one of the least palatable our imaginations could conjure up.

Another plot ripe for Hollywood would, in any case, be the revival of cryonics patients en masse. Even making the big assumption that one day such a thing would be possible, would it actually happen? In centuries to come, although a few individuals might be brought back as medical freaks, psychological specimens or first-hand historical sources, it is hard to see any point in such a wholesale exercise. The world is already overburdened and its resources stretched; the population is teeming – why wake up a load of once sick, mainly old people from long ago?

Nonetheless, if we suspend any logical reasoning for a moment, it is intriguing to contemplate what it might be like if such individuals from different times in the past were, after all, brought back as a matter of routine. Each would bring the culture and ideas of different times and places, and as they established a significant presence within the population no one could assume a common knowledge-base any longer. On the other hand, we have seen that no one will need to know anything anyway, but could access databases on a just-in-time basis; the problem therefore would be essentially one of cultural adjustment for the newcomers. Another issue, however, of most relevance would be whether, when the person died once more, they could have another spell of freezing – or would they live for ever? If they are to be immortal, the planet will soon be overpopulated. However you look at it, from scientific viability through to political, sociological and economic implications, effective time travel with cryonics is about as likely as, and less attractive than, travelling faster than the speed of light.

There is another far-fetched consequence of nanotechnology, though arguably slightly more realistic. Because machines could be self-replicating, they would cost very little. The downside, recognized by even the most ardent nanotechnologist, is that if the whole process got out of hand, the world might end up covered in a 'grey goo' of self-replicating material that choked everything else. It was just this type of scenario that featured in Michael Crichton's 2002 novel *Prey*. In reality, although a nanorobot working on its own would take some 19 million years to make one ounce of matter, self-replication could speed things up. For example, if a billion atoms were doing the job, then one second would see the production of some fifty kilograms of matter!

There were those at the beginning of the 21st century, such as chemist Richard Smalley, who never needed to resort to science-fiction paranoia in order to point out the disadvantages and problems associated with the notion of self-assembling nanodevices. In an ordinary chemical reaction there are some five to fifteen atoms near the reaction site that work in a three-dimensional operation, with no more than a nanometre on each side. The big problem is that any nanorobot would have to control not just one atom but all the existing atoms in the region of the reaction. But since the robot itself is of an irreducible size, being itself made of atoms, it has 'fat fingers' that will hamper appropriate manipulation. There is simply not enough room in the nanospace to accommodate all nanofingers and have complete control of the chemistry.

And not only are these fingers fat, they are also sticky: because of the nature of the bonds that make up a molecule, it will be hard to differentiate, and then release, an atom for construction from a finger atom, once the appropriate manoeuvre is completed. Now add to this insuperable difficulty the ever-prominent problems of miniaturization – we have already seen that surfaces become all-important – and with them many other inherent problems including those posed by friction and sticking and it is not surprising that as yet a self-replicating machine has never been built on any scale.

Advocates of nanotechnology argue that it is feasible by drawing analogies with nanomachines that already exist in nature; they point to the process whereby a genetic instruction is translated into the manufacture of a protein via RNA, or the conversion of light into energy in plants via the chloroplast. Another example is the baton-passing of electrons within the erstwhile bugs, mitochondria, that in more recent evolutionary times have moved into cells where they convert oxygen into energy. Now compare the artificial nano-counterparts: even though their small size would require only a small amount of power, it would have to come from somewhere. A more serious problem still is the amount of information a self-replicating machine would need to make itself, to collect from the environment all materials necessary for energy and fabrication, and to assemble unaided all the pieces necessary to make a copy of itself. To answer these needs in biological systems DNA and mitochondria come to the rescue; but it is less obvious where to start without these mainstays of living cells. As a chemist, you ponder whether there is any nanotech-nological device made that is independent of DNA that could ever really be more efficient. And in any case, what would be the point of trying to improve on evolution?

In your view, a more valuable use of nanotechnology is for it to assist biological, DNA-based systems in new medical applications. For example, nanoparticles can now deliver to specifically targeted sites, including places that standard drugs do not reach easily. Another possibility could be to reveal what sets of genes are active under certain conditions, for example with gold nanoparticles bound to a DNA-probe: if the DNA in question is at work they will bind, and a colour change will occur. Another device might enable quick diagnostic screens: antibodies labelled with magnetic nanoparticles now exposed to a brief magnetic field respond with a strong magnetic signal if they are reacting with certain substances. And gold nanoshells linked to specific antibodies that target tumours could, when hit by infrared light, heat up enough to destroy growths selectively.

Additional biomedical applications of nanoscience are starting to mushroom, for instance, substances that massively enhance key tissue for non-invasive imaging of the brain or body. Another advance has been nanoscale modifications of implant surfaces to improve durability within the body as well as biocompatibility, so now you could be fitted with an artificial hip coated with nanoparticles that could bond to surrounding bone more tightly, avoiding loosening. Another idea is that of a dendrimer, a kind of coat hanger about the size of a protein but which does not come apart or unfold as proteins do, therefore allowing for stronger chemical bonds. These nanoscale coat hangers could act as a means of delivering new types of drugs.

These realistic applications of nanotechnology differ from that of the 'grey-goo' nightmare in that they exploit the small size of substances without challenging the non-negotiable principles of basic chemistry, or depending on solving the tricky riddles of energy or autonomy of agenda. The wilder aspirations of nanoscience assume that nano-devices will operate *independently* of the macroworld, and therein lies the problem. For the moment, however, nanoscience is increasingly integrated with the power and purpose of macrosystems, both biological and cybernetic, so it has truly turned into 'the manufacturing technology of the 21st century'.

Perhaps Haldane would have approved of nanotechnology as a satisfactory approach to his second question, on control 'of matter as such', not least because it could also have bearings on his third big challenge to the scientific imagination concerning man's control 'of his own body and those of other living beings'. But it's time to leave the chemist of the future, perhaps in a more focused yet less speculative frame of mind than the physicist, and see how the biomedical scientists might perform in this third scientific arena – that of the body.

A rather low-tech approach is currently taking off in 21st-century science that is conceptually innovative nonetheless in that it brings previously disparate branches of the biomedical sciences together: psychoneuroimmunology (PNI). PNI is based on the reasonable assumption that the three great control networks of the body – the immune, the nervous and the endocrine (hormonal) systems – are all interlinked. This arrangement makes intuitive sense, otherwise surely there would be biological anarchy. Moreover, we have known for a long time that any one of these systems can influence the other two. Dr Esther Sternberg, of the US National Institute for Mental Health, has pointed out that: 'These notions that emotions have something to do with disease – that stress can make you sick, that believing can make you well – all of that has been around for thousands of years, embedded in the popular culture. And until very recently, we haven't had the scientific tools to prove these connections in a rigorous, scientific way.'

That said, the medical profession in general, and PNI practitioners in particular, are amassing impressive but dispassionate documentation on the effects of stress on health; for example, a sample of hysterectomy patients at Guy's and St Thomas's Hospitals were played suggestive cassettes on the benefits of relaxing. Half of these patients were released one day after removal of their stitches, whilst of the rest of the patients, who had not listened to the cassettes, only 10 per cent were allowed home at that stage. It was a maxim of even late-20th-century collective wisdom that individuals reacting most to stress are at the highest risk of medical and mental illness. In fact, stress is a better predictor of heart disease than smoking or diet. One study tested the effects of behavioural, non-drug therapy for encouraging relaxation on some fifty patients receiving medication for hypertension; 59 per cent came off their medication altogether, and 35 per cent cut their intake by half. Meanwhile, melanoma patients receiving emotional support along with regular therapy have a staggering 60 per cent more immune cells within seven weeks than a group of patients receiving the therapy alone!

As well as the widely documented effects of stress on health testifying to the robust and real link between mind and body, or more specifically between nervous and immune systems, there is the placebo effect. Taking its name from the Latin for 'I shall please', this well-known phenomenon consists of a significant improvement in health in the absence of any overt, active, external cause or substance. For example, as far back as forty years ago cardiologist Leonard Cobb showed that 'sham' surgery, in which he anaesthetized the patient and made appropriate incisions but *didn't* tie off two arteries to increase blood flow to the heart for angina, was just as successful as real surgery.

No one is really surprised by this type of report. Everyone, patients and doctors alike, recognizes the placebo effect, so it was something of a shock when a recent study from Denmark came up with a counter-claim. In a meta-analysis, the authors of the study examined 130 different projects comparing placebos with untreated controls; the stark conclusion was that there was no difference after all, no placebo effect. But closer inspection showed that the survey included conditions such as genital herpes and anaemia following surgery, which might not be expected to be influenced by a placebo. The authors themselves admitted that if they considered only subjective conditions such as depression, then the placebo effect *is* actually significant. In fact, only recently a drug trial of a new antidepressant medication was halted because its beneficial effects were not proving any stronger than those of the placebo.

And the notion of 'subjective' illness might be much broader than we think, where we include a factor, at least, that is 'just psychological'. For example take Parkinson's disease, a neurological disorder where the radical loss of key brain cells that produce the transmitter dopamine results in an inability to generate movement – the problem is the real death of real neurons. Yet surprisingly, patients taking a placebo drug have displayed as much increased dopamine availability in their brains as those taking a drug that works directly on the neuronal chemistry to produce that effect. In another experiment the very real pain from wisdom-tooth extraction was relieved as much by fake ultrasound as by the real treatment, so long as both the patient *and* the therapist thought that the machine was on.

These kinds of studies show that we cannot really draw a line between 'objective' and 'subjective' illnesses. Of course no amount of wishful thinking will mend a broken leg, but for many conditions, such as pain or even Parkinson's disease, the state of mind can certainly be an important factor. What is the all-important defining factor? Pain and Parkinsonism and certainly depression are, unlike a broken leg, intimately linked with the ongoing state of the brain. So what might be the actual mechanism whereby our neurons can exert such a powerful influence?

One idea is that, just as Pavlov's famous dogs salivated upon hearing the sound of a bell that they associated with food, so the immune system could be conditioned. We now know that a rat's immune system can fail if conditioned to do so at the presentation of an otherwise harmless stimulus. Death could even be caused just by the sound of an otherwise neutral bell; perhaps such might be the basis for the effects of 'the evil eye' in the past. In any event, in humans, psychologist Angela

Clow has demonstrated that negative and positive mood manipulations can have immediate but different effects on the immune system. Within minutes of smelling chocolate, for example, there is a measurable increase in the saliva of the chemical secretory immunoglobulin A (sIgA), a sign that immune-system function has improved. One immediate, obvious and cheap application for the future would surely be to pipe the smell of chocolate through hospital wards! And as we understand more about which smells under which conditions can manipulate which aspects of our immune system, increasingly our environments may be set up to maximize health, and with it state of mind.

On a less prosaic level, it is perhaps only in this century that science and the medical profession will begin to take the placebo effect seriously, along with the new interdisciplinary approach of PNI. PNI might well end up contributing not just to 'tool-driven' science, by enabling us to understand the link between the immune and nervous systems. In addition, along with the clinical value of developing novel ways to alleviate suffering, this new branch of science holds great promise for furnishing more conceptual insight into the actual physical mechanisms that enable our thoughts to influence our bodies, indeed eventually into the physical basis of thought itself.

Already one study of placebos gives a hint of how the brain might be working in conjunction with the immune system. Subjects were given mild pain by a blood-pressure monitoring cuff applied to the arm. Not surprisingly this painful effect was abolished with morphine, and likewise with a placebo. However, the interesting point of the experiment was that, like the morphine itself, the placebo effect was cancelled out by a drug that blocks morphine, naloxone. Morphine works because there is a naturally occurring morphine-like hormone in our bodies, which the drug impersonates. The blocker naloxone will therefore act on this naturally occurring system; interestingly, it also blocks the placebo effect. The obvious interpretation of this result is that our naturally occurring morphine system, the enkephalins, mediate the placebo effect too.

The results of a completely different type of study also fit this idea, that giving a placebo is not the same as giving nothing. Peak relief from pain typically comes an hour after giving the inert substance, as it would for a real painkiller. If the placebo effect were merely an effect of doing nothing, we would expect a more random time course of relief. The parallel time course of placebo response with that following painkillers suggests a final common conduit: the enkephalins, via drug simulation or via a process that we need to discover involving 'thought'. A thought is, after all, just a neuronal event in the physical brain, albeit currently an unidentified one. But it seems reasonable that, whatever the details of the configuration of neurons and their mode of operation, a thought in a placebo situation would amount finally to a group of such cells releasing their own enkephalins into the body.

The enkephalins belong to a particular class of chemicals in the body – peptides. Peptides are interesting because they could work trilingually, as hormones, as messenger molecules in the immune system and as transmitters in the brain: they could work on target cells by binding to special molecular targets (receptors) present in all three systems. The peptides would be ideally suited as intermediaries, allowing the three control systems of the body to communicate with each other. If the peptides do indeed act as molecular go-betweens, they could be released during conditioning experiments, or in real life according to certain states of mind. These states of mind arise from some kind of changing configuration in the neuronal landscape of the brain, where something, some holistic aspect of brain function, is 'read out' to the endocrine and immune systems; indeed read-out to the vital organs and rest of the body, via the peptides, would generate a cohesion between mind and body – a cohesion we call 'feeling anxious' or 'being happy' or even just 'conscious'.

Until as late as the 1980s scientists didn't really take the 'problem' of consciousness very seriously. It was, after all, the kind of subject over which philosophers wrangled; no one could really define it and worst of all it was, and still is of course, an utterly subjective experience. As such, consciousness has so far proved too slippery for the machinery of objective scientific methodology; one recent quip was that the whole area was, for scientists, a 'CLM' – a career-limiting move. But gradually things started to change. Very senior scientists who no longer needed to fret over the progress of their careers – a surprisingly large proportion of them Nobel laureates – started to tackle this fascinating issue: how might the physical brain, so banal in aspect, so unexciting in texture and colour, render to some mysterious inner observer, some disembodied 'you', the first-hand experience into which no one else can hack?

Over these last few decades some scientists have looked up from the bench to stare out of the window, but still no single area of research has, or can claim, a monopoly on how to go about tackling this problem. Actually, scientists like myself, we neurobiologists who deal with the actual nitty-gritty of neurons, have been least conspicuous in contributing to the debate. But, confronted with the physical, inert brain on the one hand, and the elusive, intangible feel of being 'you' on the other, one of the biggest questions remaining for scientists of the future will be: how can we start to gain purchase on understanding the process by which the 'water' of neurons is turned into the 'wine' of subjective experience? This has become known as the 'hard problem'. It will be a central question for 21st-century scientists.

By the time the next generation or two of scientists inherits the hard problem it will be possible to range easily across different subject areas, cross-referencing and correlating different types of phenomena as part of the new 'neuroinformatics'. What type of areas would they access? To start with the most macro-stage of all: the actual behaviour of the whole person in their environment. This is the approach that, among others, the evolutionary neurophysiologist William Calvin, cognitive psychologist and philosopher Dan Dennett and linguistic psychologist Steven Pinker are all currently adopting. In each case the evolutionary angle, and along with it consideration of how human thought and behaviour are different from that of the rest of the animal kingdom, can throw up enormously valuable insights into issues such as how genes differentially contribute in different species to respective behavioural repertoires. Moreover, we can start to appreciate just what kinds of behaviours and thoughts make us humans so special. But when it comes to consciousness itself, the kind of question we can ask relates not to the hard problem, the essence of consciousness, but instead to an ancillary problem, its evolutionary value: 'Why and when did consciousness evolve?' Or put even more bluntly: 'What is the point of consciousness?'

There is no doubting that the survival value of consciousness is worth thinking about. A favourite ploy is to try to work out what exactly you would *not* be able to do if you were not conscious but nonetheless the machine you are – with inputs from the senses and outputs of muscular contractions, but without any private inner world going on in between. It's actually virtually impossible to think immediately of any great loss, in terms of what actions and reactions would occur in the outside environment, but then we are talking about behaviours – and that is precisely the point. Consciousness is not a behaviour. It is an inner state, a subjective experience that can be dissociated from outward actions, the physical contractions of muscle. After all, someone lazing in the sun may not be moving at all, but is still conscious; by contrast there are numerous examples, as we saw in Chapter 3, of systems that are capable of sophisticated responses but are without any hinterland inner state. So it's pointless to approach the value of consciousness from the issue of behaviour.

Perhaps instead we should start with consciousness and simply ask what it does for us, how it enhances our lives: the short and obvious answer is that it makes life worth living. Recall that when Tony Bland, the tragic victim of crushing in the Hillsborough football stadium, was deemed incapable of regaining consciousness, the courts eventually allowed food to be withheld until he died. The message here, then, is surely if you are not conscious, you might as well be dead. We survive to be conscious, rather than use consciousness as some optional extra to survival.

What happens if we explore on a scale that is smaller and thus more detailed: the actual brain itself? This is the territory of the neuropsychologist. The aim is to explain impairments in consciousness, caused by brain damage or a contrived experimental paradigm, or both, to gain insight into how brain mechanisms might normally underpin consciousness. Using ingeniously devised tests, the skilled neuropsychologist can home in on the essence of the deficit, see how it compares with the known impairments of classic syndromes and, from there, extrapolate how the area of brain damage might match up to the problems or paradoxes being observed. Perhaps the most famous example of this approach is the definitively paradoxical-sounding blindsight.

Sometimes a person can suffer damage to part of their brain relating to vision without necessarily being completely blind. Instead, they might have just a patch of blindness (a scotoma) that means that part of their visual world is black. For the last few decades, neuropsychologists have been fascinated by the fact that if an object is aligned so that it coincides with the black patch in a patient's vision, that patient will understandably claim that they cannot consciously see it, but they can point to it with an accuracy well above chance. The excitement for the scientists is that here is a situation where consciousness can apparently be dissociated from the subconscious workings of the brain. Moreover, the effect can be studied whilst imaging the brain and identifying the critical brain regions that are or, more tellingly, are not working.

But although this type of study seems to offer an answer, the question is not so obvious. The philosopher John Searle has pointed out that when a blindsight patient tries to see something in their blind spot and fails, they are still just as conscious as they have been all along. So the central issue here, then, is about different conscious experiences, rather than the fundamental and more difficult question of how *any* subjective experience is generated in the first place.

And the use of brain imaging as a technique for answering such a question can give us the impression of understanding more than we actually do. As we have seen previously, all any imaging technique will achieve is identification of the bits of the brain that are active and hard-working under certain experimental conditions. We can go a little further and see the difference in the configuration of brain regions when we are paying attention or not, or, as in the case of blindsight, when we admit that we cannot see something. But it does not follow that the area that then fails to light up is the 'centre for consciousness'. Rather, that brain area will be playing some part in shifts of attention, not in the critical yet elusive physiological process that differentiates our brains from all other biological systems and synthetic objects and devices in the world. Indeed, if we do deprive the brain of consciousness, with anaesthesia, then multiple areas will shut down activity, not just one. In any case, the imaging techniques currently in use are incapable of showing the brain at work over the sub-second timescale over which consciousness occurs and at the same time displaying the dynamics of the shifting configurations of brain networks. And even if such a feat were possible, as it may be in the future, it would tell us only what was happening and where, not *how* a physical snapshot could be translated into a subjective sensation.

So let's push on down to the more fine-grained level of neurons and neuronal networks. In the laboratory using indirect and invasive techniques we can readily study, as those interested in learning and memory do, how networks of brain cells respond to outside events, to experience; yet even then we will not be addressing the central question of the generation of consciousness. Nonetheless, my own view is that it is at this level, of transient assemblies of neurons, that there is most opportunity for eventually finding out more about the physical basis of consciousness. After all, we have seen that genes merely and inter-mittently make proteins, and as such contribute as a tool to the material fabric of the brain rather than setting the whole agenda. And just as there is no 'gene for' a particular mental function – least of all consciousness – so there is no 'transmitter for' a feeling such as happi-ness, nor a brain region that acts independently as a mini-brain. And if there were, we would be no wiser in elucidating the hard problem – we would only have miniaturized it.

The main difficulty that we as scientists have with consciousness is that quintessential subjective quality of first-hand personal experience that makes it so hard to pin down in the physical brain. Such an elusive, purely *qualitative* phenomenon is, of course, not tractable to the way of things in a lab. One of the most dreary-sounding descriptions of a scientist, which may well have turned off many a schoolchild, is that we are in the business of 'measuring things'. Yet far from simply peering at read-outs of different data about solids, liquids and gases, what we are really dealing with are phenomena or reactions or processes that vary in degree from one situation to another, for one reason or another. What if consciousness was not all or none, not some ineffable, magic quality of the brain, but something that also varied in degree?

Scientists of the future will probably need to search therefore for something in the brain that reflects its holistic mode of operation, at all levels from the evolutionary through to the quantum mechanical; moreover, this something will not only be describable in both macro- and also microscale brain processes but will also visibly contract or expand to correlate with varying degrees of consciousness. We can, then, think of consciousness as a phenomenon that deepens or lightens, expands or contracts, is more or less from one moment to the next; it would be a phenomenon that is essentially variable and ranging in quantity from the here and now, the 'booming, buzzing confusion' of an infant or the flimsiness of a dream or a drunken moment to the deep self-consciousness of introspection of the adult human. We could then see how such ever-changing levels of consciousness match up with an appropriately changing landscape in the brain. But what might the something be, that we could measure, that was ever changing in the brain?

The best candidate, capable of such mercurial fluctuation, is not a brain region, nor a gene, nor even a particular chemical but rather ever-dynamic assemblies of neurons. We saw earlier that brain cells can muster and disband in their tens of millions over a fraction of a second. Perhaps in the future there might be brain-imaging techniques that could capture in a split second the precise formation and subsequent break-up of large-scale working assemblies of neurons throughout the brain. Only then will scientists be able to see how good a physical correlate of consciousness such assemblies actually are and only then – finally – will they be able to explore the mechanics of their formation as a critical step towards understanding how the brain generates the subjective state.

At this fine-grained level, the mathematician Roger Penrose, together with the anaesthetist Stuart Hameroff and neurochemist Nancy Woolf, have attempted to work out how such assemblies might form to give the type of flash-flood coherence that could cater for a moment of consciousness. This particular approach uses the principles of quantum mechanics to show that a network of neurons could work as one 'hyperneuron', each neuron resonating with all the rest in a kind of collective neuronal operation referred to as 'quantum coherence'.

The immediate problem with this strategy is that quantum events are usually possible only in very cold environments as we saw for the development of quantum computers. However, many physicists suggest that the brain might be a special case with a special means of isolating and protecting crucial quantum events against the backdrop of our very hot heads. A more crucial and still unresolved issue is how exactly quantum events could explain how tens of millions of brain cells work together for a fraction of a second using their delicate branches to receive and send electrical and chemical signals. As it stands, all quantum-level attempts at describing this neuronal coherence could just as easily be applied to any group of cells exposed to pulses of chemicals, in the heart or kidney or in virtually any other body location. And yet hearts and kidneys are not as relevant to consciousness as brains are. The scientists of the future who are working on consciousness will have to discover what additional constraining conditions there are in brains that are not in hearts. There must be something unique about assemblies of neurons forming in the brain that needs more than microtubules and the ability to synchronize.

Moreover, were someone to come up with a quantum mechanical explanation integrated with macrophysiology it is vital to remember that, even though assemblies of neurons might well be a faithful index of degree of consciousness, in themselves such an assembly of synchronized cells will not be intrinsically capable of generating an inner state; on its own, an index of a process is not the same as the process itself. I would not mind betting serious sums of money – although it could not actually be proved of course – that a brain slice kept alive in a dish is definitely not conscious. It is not simply because of the paltry numbers of neurons available. A mollusc, for example, may well be conscious to some degree; the holistic organization of its nervous system is intact, and also connected still to the rest of its body. In a brain slice, however, much of the three-dimensional connectivity has been destroyed, and it is deprived of any inputs from or outputs to the rest of the organism.

The two situations are therefore very different: in the brain slice, a neuronal assembly – if it is intact at all – has been isolated from the biological system with which it interacts, whilst in the mollusc it would be operating in its natural context. In the former the assembly would therefore be an index with no read-out to any larger encompassing organization, whilst in the mollusc the assembly size could interface with the rest of the body, which would be intact and as cohesive as normal: in ways that are still a mystery there would be a modicum of consciousness. Future scientists would have to question how a changing network of neurons in a brain can read out the prevailing configuration, the neuronal landscape, via streams of peptides continuously reporting to the immune and endocrine systems, to the heart and lungs and kidneys, thereby orchestrating a cohesive state throughout the whole body.

In the future it may be possible to image such assembly formation in real time within the brain, and to monitor the concentrations of different peptides in the body at different times; it may even be possible to test the theory that such assemblies are indeed an index of different degrees of consciousness – but it still does *not* follow that we will be able to solve the hard problem of how the physical brain has triggered a state that translates into a certain subjective experience.

Even from this brief overview of the levels of brain organization and function over which scientists will need to range impartially we can see that a huge block to advancing our understanding of consciousness has to be the great diversity of approach, from the quantal through to the evolutionary. The really big difficulty is that different techniques and the very nature of such different types of investigation have driven the type of question that can be asked, and have provided respective 'solutions' that do not really address the hard problem itself. But in the future, neuroinformatics might come to the rescue, at least by preventing side issues specific to any one technique or discipline clouding the real, deep question. And if a new theory ensues, by garnering data across disciplines, then I suggest the following criteria as the minimum kit by which the success of competing theories from different levels might be compared and evaluated:

1. What is the question that the theory is actually addressing? If it is not the hard problem, or some variant, then the theory is not really getting to the nub of the issue.

2. Can the theory explain how the consciousness of dreaming is both the same as and different from normal consciousness? There is no doubt that the consciousness of dreams is very different from that when we are awake, assuming of course that we are not on drugs and are not psychotic. Yet various markers like EEG patterns and protein-synthesis rates in dreaming are similar to those in wakefulness. So what physical substrate will reflect the difference in the subjective states?

3. Can the theory explain how non-human animal consciousness is the same, and how it is different from human consciousness?

(This question is a variant of 2.) Most would concede that most mammals are conscious, though some suggest drawing a line beyond which animals like the lobster are mere automata. But in the animal kingdom there is no corresponding clear-cut anatomical or physiological divide. In fact, the nervous system is designed along broadly common principles from the most primitive sea slug right through to primates.

4. How does the theory differentiate self-consciousness, sub-consciousness and unconsciousness from consciousness? (Again, this issue follows from questions 2 and 3.) Though all non-human animals may be conscious, it seems a fair assumption that, like small infants, they do not enjoy the experience we call self-consciousness. And sometimes, in extreme situations of sport, dance, sex or drugs, we too can 'let ourselves go', abandon self-consciousness whilst still being conscious. Any good theory should be able to offer a description of what might be happening in the brain when you 'blow your mind'.

5. Does the theory attempt to describe how consciousness relates to the body as the boundary of self? If consciousness is generated in the brain, then a credible theory should be able to account for why we feel our bodies are the boundaries of ourselves. Although this issue might seem obvious, it will be critical in a far more mentally networked society; we will need to evaluate the dangers or absurdities of feeling part of a greater collective that breaches the fire wall of our sense of individuality.

6. Can the theory explain how the same type of electrical signals arriving in different parts of the structurally more-or-less similar cortex translate so distinctly into experiences of sound, hearing and touch? Once more, this point is far less obvious than it might seem. To recap from Chapter 3, the outer layer of the brain, the cortex, has a similar neuronal architecture throughout; we know that when certain parts are active they match up with an experience of vision, other parts hearing, and so on. Although the answer might simply be that the difference comes from the different inputs

to each zone of cortex, from the retina, cochlea and so on, the tricky riddle remains that those inputs all use the same system of electrical signalling. There is nothing intrinsic to the inputs to the visual or auditory cortex therefore that could match up easily to the difference in experience. If a theory could make progress on this point, it would be a big step towards understanding the causal relationship between brain events and subjective states.

7. Can the theory explain how different drugs, such as morphine, LSD and amphetamine, produce different states of consciousness? If we accept that there is no subcomponent of the brain, be it anatomical structure, gene or chemical, that is not only necessary but actually sufficient for consciousness, then we need to under-stand how varying the availability of different chemicals by taking drugs can give rise to different types of consciousness. The drugs do something to chemical systems in the brain, which in turn change its holistic organization and way of operating to give rise to a shift in subjective sensation. Any theory that ignores this process, or cannot accommodate it, is not accounting with suf-ficient accuracy for how the brain generates the different 'feel' of different types of consciousness.

8. Can the theory account for the effects of a placebo, and the psychological effects of certain peripherally acting drugs, such as the anti-hypertensive drug propanolol, as well as explaining the change in emotion caused by a cognitive stimulus, such as good or bad news? This question touches on the mechanism by which subjective states, generated by the brain, are triggered by indirect factors such as feedback from the rest of the body or from the outside world. There must be some intermediary change in neu-ronal networking that in turn influences how the brain will con-figure to generate the shift in consciousness. But as yet we have no idea how a succession of such events might unfold – nor even, indeed, what they are.

9. In this particular theory, what salient feature of consciousness is being modelled, and what left out? A model involves extracting

the salient feature of a system and disregarding everything else. We saw that, in the case of silicon models, if no one knew what to leave out in the first place, there would be a risk that the all-important factor for sufficiency, as opposed to mere necessity, would be omitted. If, to be on the safe side, everything were retained, then the model would not be a model.

10. What is the experiment that would test the theory, and what would constitute a persuasive 'solution'? This question is almost as hard to answer as the hard problem itself. As yet no one has put forward the *type* of explanation that they would expect if someone were to claim that they had some concept of how the physical brain gives you the inner world that only you can experience at first hand, and how it might be proved. More intriguing still is to assume that somehow the hard problem had been solved – what would be the consequences? True understanding would give the ability not just to monitor but also to manipulate. We would therefore be able, with deadly accuracy, to transform individual consciousness, hack into it and share it around.

Perhaps, unlike all of us today, scientists working on consciousness in the future will be able to deal with this list of questions; but more important still is not whether they will be able to think of an appropriate model or hypothesis to explain this biggest of questions left to science but rather whether they will be able to design an experiment that could falsify it. After all, as the philosopher Karl Popper famously stated, the whole point of science is to be able to test, to falsify ideas with empirical data.

The three big questions that Haldane posed, echoed by the cynic John Horgan, concern time and space, then matter, then our own bodies. They still stand as great challenges to the scientists of the future. If scientists as such still exist, and if human imagination and curiosity can survive the suffocation of continuous sensuality and easy access to anything in life, from a fact to a relationship, then that same awesome technology could be harnessed to stimulate new revolutions of ideas and, in turn, new technologies not just decades but centuries hence. Might someone in the foreseeable future be writing a book about the implications of the fact that space, time, matter and individual subjective experience no longer have any meaning? Of course neither such a book, nor such a someone, could exist – by definition – in such a scenario . . .

But there remains Haldane's last big question, regarding 'the dark and evil elements in [man's] own soul'. That great thinker of the 20th century Bertrand Russell foresaw the limitations of science in replacing religion, of not necessarily giving people what they wanted, and hence of contributing to, rather than alleviating, further strife. In 1924 he delivered a reply to Haldane's *Daedalus*:

Science has not given men more self-control, more kindliness, or more power of discounting their passions in deciding upon a course of action. It has given communities more power to indulge their collective passions, but, by making society more organic, it has diminished the part played by private passions. Men's collective passions are mainly evil; far the strongest of them are hatred and rivalry directed towards other groups. Therefore at present all that gives men power to indulge their collective passions is bad. That is why science threatens to cause the destruction of our civilization . . .

This paper was entitled *Icarus*, after Daedalus' son, who perished because his arrogance took him, borne on his waxen wings, too close to the sun. Just how true is Russell's black prophecy? Will 21st-century science and technology deprive us of our 'private' passions, our individual free will, and what will happen to the human tendency for 'evil'?

8

Terrorism: Shall we still have free will?

You enter an office block and accept as routine that the receptionist will hand you a perforated form for the plastic pocket of a security badge. You check in at the airport and find nothing unusual or unreasonable in having to surrender nail scissors or remove your shoes for inspection; meanwhile the sight of an unattended bag in a crowded concourse immediately attracts attention and anxious enquiries. Such scenes, if they had been prophesied several decades ago, would have been met with incredulity. But now we are no longer really at peace; security is second nature. As I am writing this, the USA has been put on 'orange alert', the second-highest level of national danger. Acts of terrorism, if sufficiently large-scale or frequent, could well put an end to civilized life as we know it – and certainly to the sophisticated high-tech existence that, for good or ill, is otherwise in store for us in the future.

Yet surely the march of technology will eventually work against the narrowing of the mind that defines the heart-and-soul fanatic. The autonomy of surfing the internet and the enhanced passivity and hedonism that seems set to characterize the mid-21st-century lifestyle, not to mention the physical distancing away from any overly persuasive acquaintances that IT affords, should all augur well for individuals without any nationalist or extremist proclivities. In fact, in the cyber-world we could adopt the virtual guise of another race and gender, so other cultures and customs would not be so alien, and we would become ever more tolerant and broad-minded. On the other hand, those same technologies might fuel the strong feelings that tribal membership can engender, with an apocalyptic dimension more destructive than has ever before been possible. An important issue in

21st-century life will be the extent to which scientific advances will magnify the potential of tools of warfare, and whether that same scientific progress will, in addition, dispose the mindset of future generations even more strongly – as Bertrand Russell predicted – towards their deployment in acts of violence.

The most obvious image that springs to mind, thanks to the series of Star Wars films and other science fiction, is the ray gun, but surprisingly it will not appear in any futuristic armoury. The use of light to project and direct energy actually began in 1958. Lasers are good for communications guidance and range-finding, but the realization of the vision of ray guns, the direct lethal application of lasers, still has very serious obstacles. In order to destroy its target such a gun would have to shoot off large amounts of energy in short bursts. This enormous draw on energy reserves would necessitate a bulky and potentially vulnerable power plant. Moreover, the turbulence of the atmosphere would pose problems for the final accuracy of laser direction.

But nanotechnology has clear potential in a large range of military uses. One idea is that supersensitive nanosensors will improve the scanning of remote landscapes like Afghanistan, whilst a very different application of nanoscience to warfare is in new types of materials for equipment; as one military man enthuses, 'with nanotechnology, we can add properties to materials that weren't there before'. By 2025 soldiers under attack will be able to change their appearance to blend in with their surroundings: nanoparticles will change the properties, such as colour, of the materials of their uniforms. In addition, helmets will be 40 to 60 per cent lighter, and tent fabric will self-repair when ripped.

Nanotechnology is currently bandied around as an umbrella term that covers the notion of merely very small devices right through to the manipulation of atoms that would defy the laws of chemistry. Inevitably, the wilder leaps of imagination create black scenarios of massive arsenals of self-assemblers churning out weapons of destruction. But this idea is not just technically difficult; it is, as we saw in the previous chapter, actually conceptually impossible. On the other hand, we should be wary of focusing our horizons exclusively on what is potentially feasible at the moment. In the future, even though the non-negotiable chemistry laws are hardly likely to be violated, there may be paradigm shifts of different kinds in military applications of fuel cells and other basic means of speeding up production that could all amount to a frightening escalation in the rate of weapon production.

But the physical sciences cannot claim a monopoly in destructive potential. Even in ancient times armies relied on the power of illness to decimate the enemy, by routinely throwing the bodies of plague victims over the walls of besieged towns. And a particularly shameful incident in British history occurred when the USA was still under the King's rule, and the occupying force aimed to quell 'disaffected tribes of Indians': the British Commander-in-Chief, Jeffrey Amherst, distributed 'gifts' of blankets contaminated with smallpox. Much more recently, in 1984, the followers of the guru Bhagwhan Shree Rajneesh contaminated salad bars with salmonella as a deliberate act of biotechnological terrorism intended to reduce voter turnout at an upcoming election; although there were no fatalities, 751 people were poisoned.

In 1995 approximately 5,000 commuters were injured on the Tokyo subway, and twelve were killed, as a result of nerve gas unleashed by a religious sect, Aum Shinrikyo. Nerve gas, originally developed in the Second World War, is not a gas as such but is either ingested as a solution or inhaled as a vapour. The toxin then works by blocking an enzyme (acetylcholinesterase) that otherwise breaks down a widespread and potent transmitter (acetylcholine) between nerve and muscle, which also operates between nerves and vital organs: the net result of the gas, therefore, is the excessive accumulation of acetylcholine at these crucial sites. Although many different processes in the brain and body depend on acetylcholine in normal amounts, the Dr Jekyll transmitter can also transform into a biochemical Mr Hyde if it is not removed from the site of the action by its special enzyme. If acetylcholine is allowed to continue stimulating its target muscles, then it is a little like pressing on the accelerator so much that the engine stalls. The muscles all over the body stop working. The results of inhalation of nerve gas include tightening of the chest with more widespread symptoms setting in, if vapour concentrations are high enough, in less than a minute. A mildly exposed casualty will experience anxiety, headache, pain behind the eyes and restlessness; a higher level of exposure can cause twitching, cramps and general muscle weakness. A victim exposed to very high levels will have slurred speech, confusion, then slip into a coma and finally die from a complete shutdown of the respiratory system. A further concern, however, is that even those surviving modest levels of exposure might still be at risk in the long term. The effects that even low concentrations of nerve gas could have on the delicate and dynamic connections between neurons, which constitute the essence of an individual, might accordingly lead to much longer-term changes in one's state of mind.

Another, very different form of chemical warfare that was also developed in the early 1940s as part of the war effort, but is only now realizing its hideous potential, is anthrax. The recent anthrax attacks in the USA, in the wake of the 9/11 tragedy, left five dead and seventeen infected. However, further exploitations of nanotechnology could eventually make agents like anthrax harmless or detect it in public areas before it could do any harm. For example, at Swinburne University of Technology, in Melbourne, artificial 'muscles' are being built that could give early warning of biological weapons.

The principle is ingenious, yet simple. The contraction of muscle needs two proteins to slide past each other: actin and myosin. Anthrax will slow down and eventually stop this sliding. In order to monitor this action, some myosin is attached to a biochip, so that any decrease in the movement of adjoining actin will register. The chip is coated with a polymer to prevent any random sticking; a laser then etches straight lines on it, enabling the myosin molecules to stick to the chip in the tracks. Now actin molecules are added. Normally they slide along the myosin tracks, but if they come to a halt, then anthrax must be present.

Anthrax is a bacterium that, when inhaled at a rate of some 2,500 spores an hour, causes respiratory distress and death a few days later. Its action is based on two enzymes (lethal factor and oedema factor) which break down the defence mechanisms of cells. However, a third component is necessary for the toxin to do its work: a protective antigen (PA) that enables the enzymes to gain entry to cells in the first place. Genetic modification of this chemical can make the toxic enzymes useless, as they cannot gain access to the target of their action; so far this mutant PA has protected rats from an otherwise potentially lethal injection of anthrax. This is only one example of how the genetic technologies are not, as they are often simplistically perceived today, automatically bad. Yet though they may be used to combat biological or chemical warfare they can also, of course, inspire a whole new range of deadly substances.

In the future as the synthetic genome becomes better understood it may be pressed into service to manufacture completely novel bugs with lethal actions. Whether artificial or modified from natural genomes, 'stealth viruses' could be covertly introduced into the genome along with 'designer' diseases. Even before molecular biologists are able to produce diseases to order they could still set in train a course of events that has disastrous consequences – by accident or by design.

For example, Australian scientists Ron Jackson, of the Commonwealth Scientific and Industrial Research Organisation, and Ian Ramshaw, of the Australian National University, in a novel attempt at pest control, recently inserted the IL-4 gene into the mousepox virus to boost rodent antibodies, making the rodents infertile. Mousepox normally has only mild ill effects but the new gene converted it into a killer disease. How easy to imagine, then, the possibility of making a similar genetic modification to human smallpox to produce a new and even more deadly form of the virus.

Even more sinister still is the prospect of exploiting particular genetic profiles of different races for selective destruction. The issue of genocide as a conspicuous factor in modern-day warfare arose in the late 20th century, for instance in Rwanda and Bosnia. Now 'ethnic cleansing' has unfortunately become a familiar term in our vocabulary. If, as current trends suggest, war in the future will no longer be fought for territory so much as for cultural or racial homogeneity, then we might expect ever more focus on ethnic targeting. Indeed, several governments may already have worked on the creation of race-selective diseases. For example, in the apartheid South Africa of the 1980s, it later emerged, scientists studied the possibility of a disease that would render only black women infertile. Fortunately, however, the creation of such conditions or diseases would be very difficult; as the Human Genome Project showed, the genetic differences between races are far slighter than they might on the surface appear.

Because of this underlying genetic similarity it is hard to see how such a strategy could be highly effective on a wide scale. Given that most of us have a complex genetic cocktail, it would be difficult to select a sufficiently conserved yet highly selective gene that could be targeted within any one race. True, there are certain societies, in Iceland or indeed the Amish community in the USA, that have already helped the study of genetic-based disease by virtue of their relative, and very unusual, genetic sameness. But even so, there is a big difference between tracking the appearance of a disorder through generations and identifying a special gene that is common to, and operative in, the majority of any one population – above all exclusive to those particular people and only thus appropriate as a target. Such ethnically selective genes, for example for sickle-cell anaemia among Afro-Caribbeans or Tay-Sachs disease within the Jewish population, occur only in a small minority and thus, from a highly cynical stance, would still only result in the deaths of relatively few if targeted.

Indeed the very opposite of such precision targeting looks to become more likely if current trends in patterns of terrorist attacks continue; a primary feature is the indiscriminate choice of victim. The terrorist aims simply to deploy chemical or biological agents that do harm. In laboratories the manipulation of many biological agents is now routine. But for bacteria there appears to be a pay-off: as one feature is improved another usually declines – as scientists enhance infectivity, for instance, it may be at the expense of the organism's ability to survive in the environment. Nevertheless we cannot afford to be too complacent that the prospect of new toxic substances will not loom in the near future. Not only might we see more virulent viruses, such as the once mild mousepox, with new genes incorporated into their DNA but we should also expect the development of more effective delivery systems, such as micro-encapsulation.

In any event, the ideal will always be to avoid contact with any toxins from the outset. Vaccination will never be a satisfactory defence, because many vaccines are only temporarily effective, and also because it is hard to know in advance what agents to protect against. Rather, not too far into the future, we shall probably see new air-filtration systems in offices and homes, along with some 21st-century version of the gas mask kept alongside the bathroom first-aid cupboard. Air quality might well be monitored, with online flashing read-out just as the temperature and time already blink from neon billboards in any city centre. Awareness of the air we breathe may soon become as routine as the sensitivity we have all developed to unattended packages – checking air quality from different monitors, perhaps even instruments on our bodies, may become as frequent and as automatic as checking the time.

Bacteria will be more on the minds of our grandchildren than bullets. Pitched battles will become increasingly expensive, relatively inefficient and pointless – as territorial disputes are subordinated to ideological ones more suited to acts of terrorism; germ warfare is a far more obvious option. One crumb of comfort is that the Biological Weapons Convention of 1972 prohibits the manufacture or use in war of toxic substances – but then again, it can only urge signatories to respect the resolve. No one nowadays, as the power of the UN is challenged, feels reassured by toothless talking. Moreover, attack with chemicals does not necessitate a diversified industrial base nor is it expensive, and above all it allows the perpetrator to escape unidentified. The same 'advantages' also apply to cyber-crime and cyber-terrorism.

A staggering two-thirds of the companies in the UK have been hit by cyber-crime in the last twelve months, the main threat coming from external hackers. Deriving as it does from computer use, cyber-terrorism in particular plays on our deep, atavistic fears of technology transforming the computer from slave into master. Yet, despite the plots that Hollywood weaves, the real controller is not, of course, a machine intent on taking over the world, nor a beefy combatant in khaki, nor a slick 007-style buccaneer but rather a Person of the Screen, a thumb-pointing, finger-quick youth. Interestingly enough, typical hackers are usually computer nerds, not necessarily of the same mindset as the terrorists. Apparently, a neutral IT-whizz is usually recruited by a fanatic, rather than the other way round. But as keyboards and screens become a feature of more and more homes throughout the world IT-agility will be a given rather than a selling point for a would-be cyber-terrorist.

The particular power of cyber-crime, and especially cyber-terrorism, lies in its linking of the physical and virtual worlds. Ultimately, it is economic disruption or physical damage to people or places that will weaken the enemy; and you can steal or destroy in the physical world more effectively and with far less risk either to your person or of being identified with a keyboard than with a bomb or a gun. For example, in a Pentagon-run exercise hackers posed as a hostile Korean force waging a cyber-war against the USA. Using only equipment available on the shelves of computer stores or on the internet, and denied any advance intelligence or security clearance, these hackers were able, within several days, to shut down power grids and 911 call-centre systems to twelve major cities, including Washington. They also gained root access to six systems at the Department of Defense. From logistical disruption the hackers created communications chaos by sending fake messages from the President and Joint Chiefs of Staff: the result was to paralyse the American military, as every order now required visual or verbal verification.

Clearly the potential for cyber-technology to offer a non-kinetic means of paralysing traditional, kinetic weapons is huge. Even conventional warfare in the Information Age needs complex planning and coordination, depending on access to massive databases. Information warfare is not really about the hardware, whether spy satellites or computers, but about changing their human operators – or, more specifically, the decisions they make. And then there are the facts themselves. A small change made to a salient fact can have devastating consequences. An astrophysicist once remarked that the worst that could befall him professionally would be for someone to alter the fifth decimal place in the constant pi: all subsequent calculations would then be flawed, and all his work useless. The mantra of the late 20th century, and the deadly reality in this one, is: information is power.

Everyone has known since classical times that false information is a powerful weapon. But the destabilization of economies and of governments through the spread of false information has now acquired a more potent feature in that it is no longer obvious. Unlike the Nazi propaganda broadcasts of Lord Haw-Haw and the traditional strategy of dropping leaflets from a plane, a communications glitch in your IT-system may be the result of a deliberate, hostile attack, the work of a lone loony hacker, or simply an accidental technical malfunction. Another possible advantage for cyber-terrorists is the fact that computer systems can be doctored to garner data from other computer systems with which they interact. Some think that this is already happening.

As terrorism supplants conventional warfare and nations are no longer safe behind fixed borders the all-pervasive cyber-world, oblivious to territorial boundaries, will prove the new battleground. Already conflict has started for real in cyberspace, or, more accurately, for virtual real. An 'e-jihad' broke out in October 2000 after Israeli hackers targeted a Hizbollah website that had been trying to incite anti-Israel violence. By the end of the month hackers on both sides had destroyed as many as thirty websites. But things could get far worse.

Barry Collin, a senior research fellow at the Institute for Security and Intelligence in California, bleakly describes how wide the scope might be to fulfil the aims of the cyber-terrorist 'to destroy, alter, acquire and retransmit' information. For example, they might plan to access the control systems of a cereal manufacturer to change the levels of iron supplement in the cereal, so that the children of a nation were killed. Another of Collin's nightmare visions is that they could place rings of computerized bombs in a city: if one bomb stops transmitting to others, then the rest go off simultaneously. Another possibility is the disruption of banks and economic systems, so that subsequently destabilized societies lost all confidence in their financial infrastructure; or the cyber-terrorist might hack into air-traffic-control systems and cause two civilian aircraft to collide. They might alter formulae at pharmaceutical factories, or change the pressure in gas lines so that a suburb detonates and burns. In short, concludes Collin, the cyber-terrorist could prevent a nation from eating, drinking, moving or living – all without warning.

But some, like Mark Pollitt from the FBI laboratory in Washington, find these black fantasies a little too hysterical. After all, the levels of supplement needed to contaminate a cereal so severely that it became harmful would soon be noticed by the suppliers; moreover it is hard to see how so much overloading of an ingredient would not change the taste, and hence come to everyone's attention. Engineering overdoses of bad-tasting additives might not, perhaps, be the best example of what could be achieved by an individual hell-bent on causing widespread suffering.

And as for sabotage of the air-traffic-control system, Pollitt thinks it inconceivable that the human element of ATC, along with all operational rules for protecting against just this kind of contingency, could be ignored. The computers in the system control nothing; they are tools of the controllers; moreover, the rules are designed to work with no ATC at all. Then again, these reassuring and moderate opinions predate 11 September 2001, after which it is harder to regard the human element as a constant force for good in anything, including directing air traffic. And more generally, human commonsense and rationality is certainly not a persuasive defence against catastrophe.

Whether the terrorist is using nanotechnology, biotechnology or IT, he or she has the colossal advantage of anonymity. In addition, cyber-terrorism, like its biological counterpart, is so effective because of our dread of the unknown: we, the potential targets, have no control over what might happen, and in any case we are selected as targets at random. If present trends in the hostile applications of the advancing technologies continue, it is hard to imagine how our daily lives will not be affected, at the very least in terms of the precautions we will have to take each day. After all, there is no point in pouring $40 billion into counter-terrorism technology, as the USA has just done, when 9/11 provided the terrible proof that terrorist acts could be perpetrated with knives and paper-cutters alone. The only way to combat terrorism is to understand it – not just to get under the skin of the terrorist, but to get inside their mind.

So, why do terrorists do what they do? After all, their ultimate aim is not to harm innocent victims for some mysterious, sadistic reason but rather to communicate a political message to a third party, invariably their real enemy. It is hard for the rest of us to identify with such a mindset, especially when terrorists are even ready to accept their own certain death. But low-tech 21st-century terrorists may not, at first glance, appear that different from their predecessors. Terrorism may seem to be an utterly modern phenomenon but its roots lie in the 1st century AD, when Jewish radicals were struggling to free Palestine from Roman rule. A contemporary historian, Josephus, described brigands (sicarii) who took their name from the small daggers they carried under their cloaks, and whose favoured method of operating was to mingle covertly in crowds and stab opponents, apparently causing the reaction of simultaneous fear and outrage that is all too familiar today. However, it was not until 1795, in the aftermath of the French Revolution, that the term 'terrorism' first became popular, to capture the fear and intimidation instilled in the general population by the newly empowered Republicans. Remarkably, the Reign of Terror was at first viewed as a positive political system, reminding citizens of the importance of virtue through instilling fear. Soon, however, it began to take on the profoundly negative connotations that it still possesses today.

Fanaticism, total fixation on one idea and the determination to carry it out at all costs, has in the past been considered a characteristic of the loner: for example, the anarchist who killed President William McKinley in Buffalo in 1901, or the Serbian terrorist who shot Archduke Ferdinand and his wife in Sarajevo in 1914, thereby ringing up the curtain on the First World War. But such single-mindedness can also be shared collectively. George W. Ball, an Under Secretary of State in the Kennedy and Johnson years, said of the defeat of the USA in Vietnam: 'What misled a group of able and dedicated men was that, in depersonalizing the war and treating it too much as an exercise in deployment of resources, we ignored the one supreme advantage possessed by the other side: the non-material element of will, of purpose, and patience, of cruel but relentless commitment to a single objective. Yet that was the secret of North Vietnamese success. A rebuke of the spirit to the logic of numbers.'

But now this disregard for logic, whether practised by an individual or collectively, is increasingly conspicuous in ethno-religious causes and single-issue terrorism, such as that perpetrated by animal-rights activists. In the UK some activists have issued death threats to scientists experimenting on animals, whilst others have destroyed fields of GM crops.

Our notion of terrorism and its subversive violence is in principle, perhaps, no different from the terrorism of the bygone anarchists in the Austro-Hungarian Empire with their hand-held bombs, yet modern terrorists are infinitely more deadly for three reasons. Firstly, they are often well funded by a shadowy network or supporting state. Secondly, contemporary terrorists may soon have access to weapons of mass destruction. Thirdly, they are even more committed to their cause than the lone assassins of the past.

Another difference now is that the line between terrorism and classical warfare is more smudged. Whilst war has primarily been an activity between armed forces, albeit with collateral civilian casualties, terrorism indiscriminately and deliberately threatens civilians. Yet around the world we are increasingly witnessing 'low intensity' conflict – terrorism and guerilla operations – in contrast to total war that takes over everyone's lives, as it did in, say, the Second World War, or the nuclear wipe-out that haunted the years of the Cold War. The coalescence of war and terrorism coincides with a blurring of the more fundamental notions of war and peace. India and Pakistan or the Israelis and Palestinians are arguably examples of communities that are neither locked in formally declared war nor at peace. And both the Indian sub-continent and the Middle East are hotbeds of terrorism.

Although some current conflicts have clearly been triggered by territorial disputes, the struggle of ideologies seems set to be the dominant issue in the future. In 1999 the President of the Czech Republic, Vaclav Havel, declared the nation-state 'a dangerous anachronism'. And strange though it might at first appear, a backward glance at history reveals just how transient, and relatively new, the concept of a nation actually is. Civilization initially established a feudal order extending from land-owning aristocrats down to bonded slave labour, a system that was eventually eroded by the mandate of growing commerce and the printing press. By the 19th century small city-states were starting to join forces, since the best scale and power base to accommodate economies, culture and communications was a nation.

And now, within the continent of Europe, the 19th-century boundaries that distinguished the then-new nations such as Italy or Germany are fading: instead, common laws and a common currency form the bedrock of a much looser and larger federal state. Within that state old principalities and dukedoms may well resurge, not as a revival of feudal governance but rather as a celebration of local dialects, customs and cuisine, which, as they always have done, give us a sense of tribal identity. The general concept of the 'polis', the city-state that characterized ancient Greece, is promising to become once again the unit of society with which we humans are most comfortable, as devolution becomes a watchword and demands for Basque and Chechen independence become more bloody, whilst the more civilized, less desperate secession of Scotland, Wales and even Cornwall from the UK becomes a more tangible dream.

But the new tribal order is not just a throwback, an atavistic exercise in ring-fencing smaller-sized territories. We live in an increasingly multicultural society; the ageing of the population in the developed world, coupled with soaring birth rates in the much poorer countries, will lead to even greater economic migration generating an admixture of religions, languages and traditions. The divisions that differentiate society in the future may well be within borders rather than across them. In many developed nations there is already a groundswell of constituencies characterized by ethnicity, religion and race. As the sense of belonging to a nation fades, affiliations to a tribe defined not by nationality but by other desiderata might become very strong. In the uneasy twilight between war and peace into which we are now drawn differences between different religions, cultures and races come into focus. If we are living increasingly cheek by jowl in societies which are heterogeneous in these conspicuous regards, then it follows that terrorism will be more appropriate than pitched battles for fighting the enemy within, their beliefs and attitudes. But in the fight for hearts and minds victory will be harder to define and recognize. So what exactly does a terrorist hope to achieve?

Acclaimed military strategist and author Ralph Peters, a retired lieutenant colonel in the US army, distinguishes between the 'practical' and the 'apocalyptic' terrorist. The practical terrorist aims for their ideology to influence mainstream thinking rather than overtake it. He or she is concerned with rights and status, not with a paradise in the afterlife; and though they may perhaps be prepared to die, such activists would rather live. The IRA and the Stern Gang are examples of this outlook, along with some single-issue activists, such as animal-rights extremists. They wish to reshape society not annihilate it, and hence there are certain lines they will not cross. By contrast, the apocalyptic terrorist is divorced from the real world, with all its compromises and shades of values, so that, as Ralph Peters eloquently puts it, 'the practical terrorist has dreams, but the apocalyptic terrorist is lost in a nightmare'; the aims and values of the apocalyptic terrorist are of the next world, not this one. The Oklahoma City bomber, Timothy McVeigh, was a practical terrorist, whilst those who perpetrated the 9/11 attacks on the World Trade Center and Washington were of the apocalyptic genre.

Though violent, neither type of terrorist is aggressive in the way that a psychopath might be. For a psychopath violence is the end rather than the means. Psychologists distinguish between 'emotional' and 'instrumental' aggression. Emotional aggression, as its name suggests, is a short-term reaction: you immediately want to lash out at someone who has hurt you. Instrumental aggression is more subtle; it is not the instinctive reaction of, say, road rage but the calculated use of aggression as a means to an end. The terrorist aims to create widespread fear and uncertainty among very large populations; aggressive acts against the innocent are the means not the ends.

How might the apocalyptic terrorists hope to achieve their ends? Their methods may be influenced by an inevitable cut-back in the future of conventional forces. Even now, science is challenging the premium traditionally placed by armies on size and firepower alone to gain advantage on the battlefield. By 2025 it will be possible to find, fix or track and target, in nearly real time, anything of consequence anywhere on earth. And in response 'stealth' technology – the art of making weapons almost invisible to the enemy – is progressing apace. 'Low-observable' paints are emerging, made of composite materials to optimize camouflage, whilst clever redesign of shape and structure now enables a two-metre missile to have the radar cross-section of a marble.

As we saw in the previous chapter, automation is increasingly taking the individual skill and human-error elements out of much of what we do, including scientific research; so everyday military routine will become ever more mindless, benefiting from the illusory simplicity of modern weapons that is actually due to their increased technical sophistication. There will be no reason why 21st-century foot soldiers should not make up a 24-hour, 3-shift fighting force, carrying on as the day ends with night vision. But why use fighters who need light to see? Brilliantly qualified instead for the potentially exhausting, non-skilled and non-thoughtful activity of fighting in the future are robots.

Robots could, at the very least, carry fuel, water, support radar or weapons systems, clear minefields and tow other robot parts or smaller robots. They could eventually even stand in for humans. Traditionally, 'walking the point' – leading a patrol – has been one of the most dangerous jobs a soldier had to do. Now a robot will take over. For example, 'Spike' and 'Gladiator', the brainchildren of RoboTrix, a company developing prototype robots for the US military, are the size of washing-machines and can travel unmanned into the most dangerous situations. The army plans that, in the future, such robots will handle all jobs that are dangerous, dirty and dull. Some predict that a truly autonomous robot is merely 'a decade away'.

However, the ultimate robo-battle, rather like the robot football game, might somehow miss the point. If you are not inflicting death, misery and fear on your enemy, then the only point of a physical fight would be to cause economic hardship as the costly machines were destroyed. But you would also be at great financial risk yourself. Even territorial gain, which will be increasingly less relevant in battles of ideology, could probably be achieved more subtly as the century unfolds. There are many other high-tech ways of striking at enemy resources than staging a pitched battle between humanoids requiring highly complex organization. Moreover, the technology for these other cyber-based strategies is way ahead of the robotics that would be required, and so much cheaper.

So, imagine greatly decreased conventional forces, because of the advent of high-tech precision apparatus; the IT-masters, who would then be key players, would be easily captured so that they, and indeed their families, could serve as hostages and human shields. The big difference between Us and Them is the sacrosanct status of the individual in our society: hence our huge disadvantage in future struggles with apocalyptic terrorists. This deep-seated discrepancy in priorities is demonstrated by the emergence of what have been dubbed 'warrior societies', in which death is considered preferable to dishonour.

Such an outlook is so alien to most of the rest of us that it is inconceivable how and why these 'warrior societies' have come to see the world in the way they have. Fanatics and fundamentalists may have rational reasons behind their deliberate actions and deeply devoted sentiments, but then so do we all; they differ from the rest of us in the degree to which they hold their convictions. In many cases, the apocalyptic terrorist is not just ready but *wants* to die. Yet, irrespective of their degree of commitment and of the lengths to which they are prepared to go, terrorists are very rarely out of their minds. Rather, their minds have been shaped in a very particular way – as human minds are by education and any and every happenstance in the wider environment. 'Terrorism is a product of its own time and place ... We don't see the process of indoctrination that terrorists go through,' writes psychologist John Horgan, of University College, Cork.

Perhaps not surprisingly, the community in which potential terrorists live is an important factor. For example, 78 per cent of Palestinians claimed in a recent poll to support the suicide-bombing campaign in Israel. The extreme or militant terrorist may be the tip of an iceberg consisting of a disaffected community giving passive acquiescence and support, to varying extents, to a prevailing culture of martyrdom and a promise of an afterlife that is painted as a literal paradise. Despite this supportive society, the prospect of belonging to a close family, effectively a cult, is an additional lure. Terrorism has more in common with cults than with organized religion, the current militant Muslim extremists being as remote from mainstream Islam as the followers of Heaven's Gate were from mainstream Christianity. Irrespective of the atrocious violence that characterizes terrorism, the key feature of such movements, as of more peaceful cults, is mind control so extreme that followers can be persuaded to commit suicide.

Professor Martha Crenshaw, from Wesleyan University in Connecticut, agrees that a key feature of terrorism, often overlooked in the aftermath of indiscriminate violence, is the cult-like perspective. 'What is important is that terrorism is typically a group phenomenon,' she says. The organization is a 'total institution': people join out of psychological need rather than political commitment. In contrast to psychiatric conditions such as depression or schizophrenia, there is no 'typical' personality type, although many of the young men who become terrorists may suffer from low self-esteem and accordingly are attracted to groups with charismatic leaders. But there is undoubtedly the unpalatable attraction, which most of us have experienced at some psychologically frail times in our lives, of relinquishing the need to act on one's own initiative. On a recent visit to China I was strolling in the sun in the massive area of Tiananmen Square, with gigantic pictures of Mao beaming down from his mausoleum and incomprehensible loudspeaker music stuttering all around and I was struck for a moment by the appeal of being at one with the rest of a massive crowd, of being accepted, and the lure of simple lack of hassle – of just being told what to do and mindlessly accepting.

There is no denying that the need to belong, to feel part of a group of people 'just like you', is very strong and deeply ingrained in human nature – from gangs in the playground to teenagers wearing the same trainers to the bar in the golf club. But what if a close-knit cyber-tribe starts to fulfil the roles of extended families or clubs; might cultural differences eventually be diluted and sanitized simply into the different ways we surf the net, no longer an incentive for hatred?

The critical issue that determines when the feel-good factor of belonging sours into superiority and intolerance might be the degree to which membership of one circle actually precludes an affiliation with any other. Most of us belong to many groups. Our loyalties are in a nested hierarchy of increasing dilution, from immediate family outwards to distant relatives, one's company or profession, religious group, sports club, geographical region and so forth. By participating in many different types of relationships we cater for all our many and diverse basic human desires. By contrast, enormous power can be exerted over someone once they are removed from such other areas of allegiance. For example, the army cuts trainees off from their previous lives so that the combat unit becomes their family, and fear of letting down comrades worse than dying. Non-violent cults, such as the Moonies, as well as terrorist groups famously use the same tactics.

So what is the lure of belonging just to one utterly exclusive group that demands so much – the entire heart and mind of an individual? A crucial aspect is the terrorist's image of those outside the movement, the enemy, as victimizers. Clearly, the terrorist must be in an oppressed minority for this idea to take hold, and irrespective of the reality of the situation the oppression must be perceived as extreme: a recurring theme of goodies and baddies. You are either on one side or the other – totally right or wrong. And once in the realm of cowboys and indians the mythical qualities of the situation exert a simple but great appeal.

'The heart fed on fantasy, grown brutal from the fare,' wrote Yeats. In quoting him (in an essay, 'The Mind of a Terrorist', included in the BBC's book on 9/11, *The Day that Shook the World*) the political journalist Fergal Keane underscores the power of mythology in shaping a collective cult consciousness. It is 'necessary to narrow the mind' to ensure the rigid adherence to a simple, single idea. In the 20th century this tactic was used, Keane says, by the Irish Christian Brothers, working with the IRA for an Ireland completely independent of Britain. Another particular, very different, example was the novel *Jud Süss* – a book initially intended as an argument against anti-Semitism – twisted at Goebbels' behest in 1940 to make a film which presented a subliminal argument in favour of ethnic cleansing. In this Nazi film the main character is a powerful 18th-century businessman from the Jewish ghetto, a hook-nosed rapist who embodies every savage caricature dreamt up by the anti-Semitic propaganda of the Third Reich. This 'bad guy' is hanged in the final scene for his crimes against the pure white Aryan women. In another move to create cult consciousness the Nazis also propagated the myth of the lost City of Atlantis, casting themselves as the descendants of a god-like race. And another basic plank to their culture of the 'warrior society' was Norse mythology featuring Odin and Thor, bellicose gods whose warlike glamour made them appropriate role models.

The romance of heroes and myths in warrior-like cults is clearly a hugely powerful factor that might have been underestimated. Osama bin Laden apparently routinely evokes nostalgic and angry memories of the medieval Crusades against the Infidel. The alleged insult still to be redressed is the massacre of Muslims by the Crusaders when they captured Jerusalem.

Whether such self-righteous, unforgiving indignation arises either in a group or in an individual, in every case the underlying drive is a thirst for revenge as a result of an attack on self-esteem. Given the extremity of feeling, and the polarization of right and wrong, a lack of empathy – an inability to put oneself in another's shoes – is perhaps not surprising. And if you know you are so utterly in the right, then clearly you must completely defeat your opponent, who is utterly wrong. You will be fighting for your psychological life: any slight or moderate deviation from the opinions of you and your comrades you will see as an insult and threat.

Given such single-mindedness, it is not surprising to find that membership of a terrorist group or cult suppresses many normal human behaviours – the different drives that E. O. Wilson described in his classic study, *On Human Nature*. The desire for sex, for example, is irrelevant compared to the over-riding mission, and indeed cult leaders often lack stable relationships. The normal, important desires for tranquillity, for order and for independence are also culturally over-ridden. By contrast, the basic human needs for power, honour and status are fuelled disproportionately – even though the status is collective rather than personal, of being completely right and yet persecuted.

Yet this is a 'chicken and egg' problem. It is impossible to say whether a personality that is already so emotionally lopsided is particularly drawn to cults or whether membership of such a group pulls and pushes the fabric of your mind out of shape. One neurological 'explanation' for the terrorist mindset is that there may be damage to an area at the front of the brain, the prefrontal cortex, which happens to be very dominant in humans – twice the size it should be for a primate of our body weight.

One idea is that the checks and balances that are normally in operation as we go about our daily lives are, for whatever reason, absent in the brain of a murderer. This has also been suggested, incidentally, about schizophrenia. In both cases, allegedly, brain scans show a malfunctioning prefrontal cortex. Yet this type of approach brings with it many problems. First, whatever the physical aberration within the neuronal whirring of the prefrontal cortex might be, it is similar for a broad spectrum of dysfunctions; but remember that terrorist thinking can be clearly distinguished from that of the psychopath – the psychopath shows immediate and direct emotional aggression whilst the terrorist uses violence as an instrument, a means to an end. A better line to pursue might be to ask what the common factor in psychopathy and terrorism could be, perhaps some kind of weakened grasp of 'reality': we could then see how such an impairment could be explained in terms of the malfunctioning prefrontal cortex.

But then we come to the second problem: how is a 'grasp of reality', be it strong or weak, realized in the interstices of the physical brain? The prefrontal cortex is related to working memory – the ability humans have in particular abundance to keep salient rules in play at any one moment. But this cannot be its sole 'function': this same brain region is hyperactive in depression, underactive in schizophrenia. Moreover, damage here can lead to source amnesia, not memory loss as such but the loss of time and space frames of reference for a particular event. Taken altogether then, the prefrontal cortex is not a mini-brain in its own right, not the 'centre for' any particular trait; rather neurons project from there as input to the rest of the brain, to drive the landscape of global brain connections into a configuration that in some way enables conscious access to a myriad of associations. It is these associations, this personalization of the brain, that enable us to have a stable and consistent view of the world – a view that is absent or unstable in the otherwise very different conditions of infancy, schizophrenia and indeed dreaming. Yet arguably all three, infancy, schizophrenia and dreaming, though very different, are examples of the same scenario, in which the constraints of the adult brain, garnered through daily life, are, for different reasons, not available. We cannot explain terrorism therefore in any meaningful way, by recourse simply to the prefrontal cortex.

Now we come to a third difficulty: even if a terrorist willingly volunteered to have his or her brain scanned, and even if certain brain regions were active in a pattern different from that seen in everyone else, it would still be hard to speak of a neurology of terrorism. Brain scans, as we saw in Chapter 3, give only a very crude picture at the moment. Just like the very first cameras, whose exposures were too long to capture movement – hence the photographs were all devoid of people – current imaging is too slow to capture the sub-second coherence of tens of millions of brain cells working together for a moment of consciousness. Furthermore, even though certain configurations of brain regions might be different from those in the brains of non-terrorists, we still would not be able to tell what was happening, what the 'function' of the particular brain regions might be.

Most important of all, such a brain image would not even tell us whether the unusual brain state was indeed the root cause of the terrorist's thinking. There is still the issue of whether the abnormal neuronal landscape has led to extremist, fanatical acts or whether the act of thinking like a terrorist has shaped the ongoing physico-chemical state of the brain. In brief then, an aberrant prefrontal cortex, or any malfunctioning area lighting up on a brain scan, really tells us very little, especially about the 'cause' of a terrorist's thinking.

Nonetheless the chicken and egg dilemma is critical if we are considering when an individual is accountable for their actions. For example, in 2001 Keydrick Jordan was spared the death penalty because of a psychologically flawed upbringing, which purportedly left him brain damaged; the scans were brought into court. And even more recently the musician Peter Buck, of the rock band REM, was acquitted of a series of charges after attacking a flight attendant; his defence was that a sleeping-pill had reacted badly with the wine he had drunk, causing him to be 'not himself' – but rather what is described in legal terminology as a 'non-insane automaton'. Still, note that the term emphasizes the robot-like aspect of Buck's action, and his absence of free will.

Increasingly, people may start to point to genes, to drugs, to childhood experiences and to brain scans to explain their behaviour, as we become more and more familiar with the different biochemical bit players and layered circuitry on all levels of brain function, and how they come together in the crude snapshots that imaging currently allows. In general, we will have to confront the big issue of whether science has made us less accountable for our actions, whether the apparent determinism suggested by various scientific observations is indeed real. If so, where do we draw the line? Could bin Laden plead that faulty genes caused him to orchestrate the 9/11 attacks? If not, why should a distinction be made between him and Stephen Mobley, who, we saw in Chapter 5, is using such an argument as mitigation in his trial for murder?

In all cases we must be clear what we mean by the term 'individual', if questioning individual accountability. Although the law has long recognized the concept of *mens rea* – the state of mind indicating the culpability that is required by statute as an element of a crime – science is now challenging how easily that concept can be applied. The influence of alcohol or medication differs from a disrupted childhood or disrupted genes, affecting a state of consciousness rather than a lifelong mindset.

An obvious similarity between defences resting on alcohol or road rage is that the changed state is a short-term, acute one: consciousness tilts for a moment, and you 'lose your mind' – you do not access for that moment the appropriate associations. But the problems ensuing from genes or from childhood are longer term, chronic, reflecting the mind itself, the enduring configurations of neuronal connections. Now, we have seen that terrorists do not suffer from any conventional psychiatric disorder, and indeed that they do not have an unnatural tendency for emotional aggression, as a psychopath might. So the problem is not apparently one of a loss of mind, as in road rage; rather the terrorist's quintessence is a more enduring state of mind, one that has been, by all usual standards, twisted. The neuronal connections that make up the mind of the terrorist are not based on a single faulty gene – not least because such a thing is unlikely to exist – but rather on a complex set of dysfunctional ongoing circumstances that have made up that particular life narrative, and on the ongoing environment to which the individual is exposed.

There is no sign whatsoever that such predisposing environmental circumstances will abate in the future – in fact, given current cultural, political and economic trends, they can only be set to increase. Hardly surprising then that many think that we will see an increase in the more chilling 'apocalyptic' brand of terrorism, with no basis for rational discussion nor any restraining fear of death – either for oneself or others. In 1999 bin Laden proclaimed that 'hostility toward America is a religious duty', and it is hard to be optimistic that, with this belief, he will not go on to try to kill millions. So in the future how might terrorism be stopped?

In any 'solution' science will play a part. First, in the most obvious way, if a strong pre-emptive strike is required. Here the physical weaponry of traditional warfare will be harnessed, to greater or lesser extent under cyber-control, to cause physical damage to people and property. Secondly, more subtly, sophisticated IT will be used to develop an effective counter-intelligence strategy. However, neither of these would 'solve' the problem of terrorists who put little premium on material objects, nor indeed on life itself.

I have suggested that the problem of apocalyptic terrorism will never be completely 'solved', since we are confronting a mentality that has no reverence for individual life. Yet most of us need to believe in something more important than life: unlike all other animals we know we are going to die, so we give life and death a meaning. There are even formal experiments that show that the more humans think about their own death the more strongly they embrace the values of their culture. The terrorist mentality is surely an extreme example of this link working in reverse: embracing such strong values, death is omnipresent but less relevant.

So surely we should tackle what terrorists themselves prize, not territorial conquest or individual lives but global conquest of hearts and minds. Think of terrorists as at the top of a community pyramid of like-minded folk with a graded intensity of views. The hard-core extremists must constantly recruit and mobilize from the lower, more moderate levels of the pyramid – the all-important sympathetic community. Extremely aggressive retaliation, like the bombing of civilians in Afghanistan, will serve only to create more supporters of the grievance that fuels the terrorists' cause. Meanwhile, the biggest enemies of the terrorists are the moderates on their own side: an effective way forward might therefore be to avoid a knee-jerk reaction of stereotyping and prejudice against all moderates of any religion, race, nationality and culture similar to the terrorists, and to seek instead to divide and rule. And if we wish to marginalize the extremists, then we need to harness technology to create the kind of life that will reverse any discontent, hardship and distress that fuels the mythology of oppression. We need to give people back a sense of their own personal identity, rather than a collective one. We need to give them back a private life.

Perhaps 21st-century technology does indeed hold the promise of physical ease and sensory gratification for everyone. If you buy into the idea that the mind is the personalization of the brain, the organization of neuronal connections through experience, then that brain will be highly vulnerable to 21st-century technology. As this technology becomes more pervasive then in theory anyone anywhere in the world could start to live in a cyber-world – a world where no one is accountable, everything is determined and everyone becomes increasingly passive and reactive, perhaps losing not only the rarefied curiosity that drives science but also their own identity – individual or collective. IT may open a window onto other cultures but only if humanity has retained a mind enquiring enough to peer out.

If we had, in the future, the combined force of IT, biotech and nanotech, might we eventually be able to lull not just the moderates but the terrorists themselves into a happier frame of mind? But there would be no easy segregation of those 'deserving' to be so targeted and those able and willing to exercise a free mind, no obvious distinction between those who might be disaffected and become sympathetic to terrorists and free-thinking, broad-minded, rational individuals like ourselves. And if there were, it would be an elite of the most pernicious kind. So instead, assume that we would all be appropriately manipulated. Would we want a society, call it outcome A, in which we are all free from the fear of terrorism, 'happy' and physically very comfortable, but at the same time have abrogated personal responsibility?

A second scenario, outcome B, is the polar opposite: there is no physical comfort or freedom from mental persecution and misery. You are an individual with a brain that has escaped the intervention of modern technologies: but, as in times gone by, your mind is imprisoned within your own interior. You are unable to express to the outside world your emotional or material needs because of the constraints of the narrow-minded, uncompromising society in which you live. Such would be the result – so long as nuclear war had not erupted – of a repressive, fundamentalist state of the type promulgated by the current generation of apocalyptic terrorists.

In the short term such a dichotomy is, of course, crass and overly simplistic. Probably what will happen is that we shall continue into an ever-pervasive grey area, that uneasy world that is neither war nor peace. Our lives will become edgy and insecure, accompanied by anxiety every time we board a plane, ride an elevator to the top of a skyscraper or open a bulky envelope. It will be a life dominated by security alerts, by tragic newsflashes, by flashing warnings of air contamination, by diverse devices on our person and around the house anticipating cyber- and biological or chemical attack. Perhaps such uncertainty will be the only element of chance in an increasingly techno-controlled lifestyle. Above all, you will still have a chance of being you: outcome C. But can such tension continue? If so, surely it will change us in any event. Perhaps the perpetual fear will drive us increasingly to seek out the oblivion of drugs or a saccharine cyber-never-never-world: outcome D.

The alternatives are stark. In outcome A, you would effectively no longer have a mind of your own, though you would have the physical potential to express it, whilst in outcome B the potential for individuality would remain, though the physical external conduits would be denied you. Both outcomes represent a loss of your human potential; in the latter case you would be aware of your predicament whilst in the former you would not. In terms of your mind, outcome C is a more moderate version of outcome B, whilst outcome D is a less complete version of outcome A. So the basic question is whether or not we end up retaining any sense of individuality in a compromised environment. Is that individuality, the birthright of each human being, more robust than the external influences that could become utterly controlling; or are we facing up to the bleak prospect that for the first time in some 50,000 years human nature will be transformed?

9

Human Nature: How robust will it be?

We started our journey into the future by simply wandering around a new type of home smiling, gasping or shrugging at all the various gadgets. But from the outset the big issue has not really been the lifestyle that these clever devices will make possible; more fundamental still are the attitudes that the new technologies will engender in the late-21st-century you – how you will view reality and, most important of all, how you will see yourself. The single biggest difference between the present and the future you is that in the future you will be living in a fundamentally different world *where there are no clear categories of any type*, from the most banal physical objects to the most slippery mental concepts.

Let's start with the banal. In your home even workaday physical objects can no longer be readily classified, but change shape to suit the context of the moment. And your every movement from room to room, your inner body processes and your spoken words all influence the environment from one instant to the next. You no longer think of 'reality' as something enduring, independent and somehow 'out there' – indeed the term has drifted into obsolescence. Nor is there a clear line between the cyber-world and the atomic one. Your friends and colleagues can adopt cyber-profiles, whilst all domestic facilities have human interface personas. You can watch both human and cyber-characters on the screen, roving around your home or in virtual scenes such as the supermarket. And you can choose to impose yourself and your friends into fictional settings.

Just as it is hard to demarcate reality from fantasy, so it is difficult even to be sure about the boundary of your body: where does the physical 'you' actually end? After all, you probably have at least some synthetic prostheses to improve and heighten your senses and to boost flagging muscle power; not that you really need muscles, as you can now merely 'will' objects to move. But it is not just the non-carbon components of your body that blur the essence of 'you'. Another boundary breached is the one between physical and mental events: you are acutely aware that every thought you have is impacting on your immune and endocrine systems and on your vital organs. Thinking and feeling are now intimately merged: you are keenly aware of how your moment-to-moment emotions, and in the longer term your health, can hinge on a worry or a compliment. So now where are the previously obvious distinctions between mental and physical, true and false, objective and subjective?

In addition, like most people in modern life, you now carry a 24th chromosome with temporary genes that you will probably upgrade when you donate your gametes for reproduction. And if you decide to pass on your combined natural and artificial gene package, there is no longer a line between reproductive age and infertility, nor does it matter whether you are homo- or heterosexual: the production of progeny is, literally, boundless in possibilities.

The traditional dichotomy between work and leisure has vanished too, in terms of location: everything happens from home, and in terms of time, your day is completely fragmented, whilst both work and leisure activities are screen-based. And if you ever were to engage in a 'fact-based' pursuit, accessing the latest science or even performing a cyber-experiment, then once again you would find that old-fashioned classifications have disappeared, in this case between biomedical and physical sciences.

But most of your time, as for most people nowadays, you spend oblivious to the past or future, whiling away the moments in the twilight interface between the imagined and the actual. And in interacting with others, too, everything is ambiguous. Fictional dating partners, and indeed the entire virtual family, really do seem imbued with all the inner turmoil and emotions of real people – it is just that they are eventually compliant with your whims. The stereotyped roles associated with a bygone 'classic' relationship are all in the past. Conventional dating and courtship patterns are no longer appropriate or necessary: instead the full gamut of emotions and sensations entailed in a complete sexual relationship can be provided artificially by a combination of IT and, most important of all, your receptive mind.

And the ambiguity extends into all family relationships: there is no longer a clear mother–father–child nuclear structure. Even in the previous century the concept of the blended family had increasingly become the norm. Now many children may have as many as six parents, including the providers of gametes, host-eggs and a womb as well as those making a purely environmental contribution to child rearing. Moreover, this generational relationship is far less clear-cut than it used to be. Because children now glean much of their education, both formal and informal, from a screen that condenses all time and space, the notion of 'parental influence' is regarded as a historical phenomenon that was already on the wane at the turn of the century.

As the nuclear family vanishes, the increasing number of older people confuses the picture still further. Not only can the senior generation carry on reproducing, and be grandparents and parents simultaneously, but they are also far more agile and healthy than in the old days, so the differences between them and the younger generations are no longer so obvious. Moreover, these older people cannot be readily distinguished from their successors by what they know; everyone can now access all facts immediately. The experiences of the elderly are no different from anyone else's; for the most part life experience is second-hand, recorded, and annotated by augmented reality. Since no one actually lives life in real time anymore, or in a constant physical environment, the benefits of experience and the acquisition of wisdom to deal with unforeseen circumstances are no longer at a premium.

The nuclear family, the icon of the post-war years but already strained by the late 20th century, has now disintegrated completely in the face of ambiguous generations, complex reproductive relations and a pervasive cyber-world. The need to belong to a group, a tribe, is catered for by the virtual and real friends you contact via your screen, who are so much more compliant and more tolerant of you and your whims and habits. But then your 'place' in this society is not fixed as it used to be in previous eras – not only because society is no longer divided into families, class or generations but also because you yourself are no longer a well-defined entity.

It is almost as if the old thought experiment 'colour-blind Mary', popular with philosophers at the turn of the century, has come true – only in reverse. The utterly unlikely story of colour-blind Mary was conceived by the philosopher Frank Jackson to illustrate the difference between understanding and direct experience. Mary (at least she is female) is a brilliant scientist who understands everything there is to know about the physiology of colour vision, but she herself has never experienced it at first hand because she has been raised in an entirely monochrome environment. Now, if she is unleashed into the real world of vivid colours, will she learn anything new? The critical issue here is the question that we are asking about Mary's state. Will she learn anything more, *understand* anything more about how the visual system within the brain that processes colour? Probably not. Will she have a dramatic change in her conscious state? Almost certainly yes.

In your late-21st-century incarnation, you are a kind of Mary-in-reverse: raw, direct experience has replaced knowledge and insight. If true understanding is seeing one thing in terms of something else, of making a connection, then nowadays you understand nothing at all. In your present world no person or object or process is unambiguously and consistently linked to anything else; instead you live in a blur of pulsating sensations. You have none of Mary's understanding but all the first-hand sensation that she was missing. Your moments are visceral, sensation-laden and invariably abstract. Perhaps even your senses are not stimulated in the way that they would have been in the 20th century, but then how could you know?

As Mary's awareness perhaps changed, when she finally experienced colour, your consciousness will be changing. But for you, it will mean very little. You do not have an attention span of sufficient duration to work out an explanation, to track a plodding sequence of events or premises so that you are able to place any particular issue in question into a context of other places, people, objects or events. Most important of all, you do not have the motivation. Why should you bother to make an effort to force a connection between random bits of information? You can just access the ready-made conclusion as a stand-alone fact. You are a Person of the Screen.

So, your world is highly interactive, highly personalized and highly unstable – changing both within and around you according to whatever thoughts you might have. Yet these thoughts are disconnected, reactive to the moment, as fragmented as your life is and similarly with no narrative continuity, no meaning. 'Who are you?' or 'What are you?' – even 'Where are you?' – are all hard questions to grasp nowadays. And perhaps, if you have no delineated thoughts, ideas and body of your own, then perhaps you are not an individual at all . . .

This depersonalization is both a cause and effect of the most basic fact of modern life: for every move you make, or rather for every thought you have, there is a record. Your entire life narrative – from your daily bowel movements to the changing status of your immune system to your choice of entertainment – is logged and open to third-party scrutiny. Remember that, thanks to quantum computing, nothing is confidential any longer; every chemical, biological, medical, psychological, financial and social detail about you is public.

The bitter truth, however, is that this doesn't really matter. Since everyone is so homogeneous, in health and behaviour patterns along with reproductive and relationship options, neither you nor anyone else for that matter is that interesting anymore. Even the once individual and much vaunted genome is now eclipsed by standardized upgrades, increasingly powerful yet reducing variety; in any case the limits of gene technology, especially in relation to mental function, are now much better understood. On one hand, intervention at the level of a single gene is truly valuable for reducing the risk of a problem where the genetic profile has been established; on the other hand, no one now expects the impossible – a vague enhancement of a highly desirable and subtle mental trait achieved by tweaking a single strand of DNA in every cell of the body. But in any event there is not much demand for that kind of thing anymore; after all, what would such enhancement get you? More lovers? Children? A better appearance? A better job? None of these values are meaningful anymore; there is, quite simply, nothing left to gain from the status that they would bring. For the first time in the culture and history of humans' status is irrelevant. Experience is all there is.

So take a look at yourself living out your life narrative in front of a screen, watching yourself watching yourself. You are in no discomfort, all your bodily needs are satiated, and you can turn fantasy into a cyber-reality that seems as real as anything you have ever experienced.

All your experiences, which are for the most part virtual, are documented – and these experiences, which can't really be distinguished from your thoughts, feed into a worldwide network.

Early in the 20th century Pierre Teilhard de Chardin, a Jesuit, conceived a most bizarre and visionary notion, especially by the clipped and clear standards of the imperialist era: the concept of a 'noosphere' – from the Greek for 'mind', 'reason'. His noosphere was a collective system of thinking that linked up all individuals around the globe. Now, of course, such a scenario does not seem so crazy, substantiated in part as it has been by the web and the net. There are those who have already likened the web-net to a brain, each individual, fancifully compared to a neuron, eventually communicating incessantly with each other. But the most important aspect of de Chardin's vision has been so far overlooked: each person in the noosphere is completely subsumed by the greater, collective consciousness. In brief, the notion of an independent individual, with a private life and a unique portfolio of thoughts, knowledge and opinions, is finished.

I imagine that everything about this strange future would be abhorrent were it not happily offset, for most sensible and realistic folk, by its sheer improbability. For a start, you sigh in world-weary complacency, the technology is not available and never will be, so it just could not happen for practical reasons. The Cynics, whom we met at the very beginning of this book, were right. Further, you consider, such a situation could never really arise even if we had the technology simply because of the 'yuck factor', that redeeming reflex in all human beings to say 'enough is enough'. Human nature, being what it is, will surely damp down the excesses of the gadget-obsessed, dysfunctional nerd. A technology won't necessarily be realized just because it is possible, especially if it violates our sense of what is right. And anything that depersonalized us, made us less individual, would go against the grain of human nature.

Up until this current moment, whether we were in a medieval court or the modern Amazonian rainforest, we would always have had a sense of identity: most of us adult human beings feel our individuality very keenly most of the time. We are aware that we have a mind that is like no one else's, that we see the world in our own special way. Unless we are swept up in a strong sensual or sensory experience where we, tellingly, 'lose our minds' or 'let ourselves go' we are continuously conscious of our selves as distinct entities. This individuality is, for most of us, our most treasured asset. After all, the dystopias featuring in the two great novels of the mid 20th century are so chilling to encounter not because of any gadgets that they catalogue but precisely because they threaten the sense of self.

Aldous Huxley's *Brave New World* paints a future dominated by, as we would now see it, genetic engineering. Society is carefully stratified into levels of humans genetically destined to have different abilities. George Orwell's *Nineteen Eighty-Four* threatens individuality by breaking down the privacy of the inner world, the mind. Citizens are under almost constant surveillance, and the plot thereby anticipates the possibility of a new information technology for monitoring body and brain processes that invites not only third-party scrutiny but also, as a logical consequence, third-party manipulation.

When both these books were written biotechnology, information technology and the science of quantum theory were still in their infancy. Half a century or so later the idea of manipulating genes, and even manipulating brain processes with implants, is no longer science fiction. Many of the diverse developments that could soon be transforming all aspects of our lives are repugnant basically because their misapplication could lead eventually to the loss of the sense of self as a distinct entity.

Until now it may have been easy to think, as The Cynics still do, that simply being human, 'knowing our own mind', is the most appropriate and effective bulwark against the technology threatening to engulf us. But we have seen that our grandchildren are destined to have different fears and hopes from ours, or even none at all, engendered by different influences on their brains. The more we learn of the exquisite dynamism and sensitivity of our brain circuitry, the more the prospect of directly tampering with the personalized brain, the mind, becomes a distinct possibility. And the big issue is this: these influences have the potential to be far more pervasive and more direct than anything that has gone before, even beyond the modern solace of punching the switch of the bedside-lamp, a simple yet far-reaching invention, to instantly expunge a nightmare. At the technological limit, then, just how robust will human nature turn out to be? Before we can answer this question we need first to agree on what human nature actually is.

Until now, human nature has often been invoked to 'explain' illogical or irrational behaviour, that undefinable something about being human, sometimes against all odds. Perhaps the collapse of the Berlin Wall is one example; on a smaller scale, the personalization of a bleak cubicle in an open-plan office with postcards and photos or the increasing popularity of farmers' markets all stand testimony to 'human' requirements in our lives, beyond the technological and merely functional.

I found another example of how the David of human nature has restrained the Goliath of modern technology in a new hospital that I visited recently in Oslo. As we looked down from one of the gantries spanning a broad corridor my host explained that the whole design was meant to resemble a street. The hospital was like a village with a main street, the corridor, where staff, patients and visitors could wander, browsing in the shops and cafés along the way. Particularly sensitive to our natural disposition was the way that this corridor was deliberately designed to curve rather than run in a functional straight line – simply because streets in villages have never been dead straight. The effect was indeed remarkably reassuring and comforting in what could otherwise be perceived as a frightening, impersonal atmosphere. And let's face it, it is the impersonal and the 'depersonalizing' large-scale that scares us – which is why some of the predictions in this book tend to be so distasteful. So how might the personal, the individual-sized human factor, be described in biological terms?

Over a hundred years ago Charles Darwin propounded the theory that emotions are universal, and therefore surely part of what we could now call human nature. Following on from this ground-breaking insight the psychologist Paul Ekman, of the University of California Medical School, has much more recently classified human facial expressions as revealing any one of six basic emotions: fear, surprise, anger, happiness, disgust and sadness. One of his seminal studies was of an isolated, pre-literate Stone Age culture that had had no contact with television, radio or any other lines of communication with the modern world. Ekman showed these isolated societies photographs of people with the basic range of facial expressions, and instructed them to point to the one 'where the person is angry', and 'about to fight', and next to indicate the person who has 'just learned his child has died'. The responses were the same as they would be in our Western culture, and were the same in every society where the study was conducted.

This recognition of facial expressions, common to all humans irrespective of their culture, is clearly deeply ingrained, and occurs automatically. Ekman can detect tiny changes in the 'fight or flight' systems of the body – changes in heart rate and blood pressure, sweaty palms – when he shows his subjects the six basic emotional expressions; amazingly, these reactions occur even before the subject has consciously recognized the actual identity of the individual face.

Although these feelings are common to all adult humans, they are arguably not all exclusive to our species. True, disgust is hard to identify in animals, even in chimps, but then neither does it seem to be present at all in small children. For example, infants have no reservation about eating chocolate that in appearance and colour resembles faeces or drinking urine-coloured apple juice from a (pristine) bedpan. Once beyond a certain age, however, these behaviours are simply unacceptable and are met with, yes, disgust.

And just as animals and children are alike in not registering disgust, so both constituencies are apparently capable of other emotions in the 'universal' portfolio, such as happiness. Even a rat will work at pressing a bar, instead of eating, to stimulate electrically its brain in certain areas operationally dubbed 'pleasure centres'. Of course we will never know what someone else, let alone a member of another species, is actually feeling – but on the face of it a purring cat or tail-wagging dog is not that far removed from a gurgling, smiling baby. Similarly, the sadness that a small child might experience on losing a toy could be compared with that of a dog, listless, with tail between legs, who has lost a human or canine companion. Undoubtedly positive and negative feelings – let's not even call them happiness and sadness – can be generated in non-human brains. So we cannot simply equate basic, 'universal' emotions with *human* nature exclusively.

Some thirty years ago the biologist Edward O. Wilson argued strongly for the durability of human nature, which he defined as 'those deeply embedded laws of behaviour that shape society, technology and culture'. These traits and tendencies have conferred selective advantages, such as cooperation, spirituality and tool-making. Nothing has changed for 100,000 years and thus, he argues, nothing is likely to change in the foreseeable future. According to Wilson humans will always show a tendency towards hierarchy; personal concern for status; value for self-esteem; desire for personal privacy, including space; deep sexual and parental bonding; aversion to incest; and tribalism of some kind, even, presumably, of soccer teams. We can also add a desire for tranquillity; for independence; for power; for excitement; for a sense of order; a predisposition to laughter; and above all the need, however vague, for fulfilment. Wilson describes as the 'human condition' the sad fact that hatred and aggression coexist with a sense of ethics. Promiscuity is 'not in human genes' so, Wilson says, the family unit is secure.

The list may offer a good series of examples of human nature but we might come up with others, such as desire to personalize the workspace, or predisposition for curvy corridors. Surely more important than anything, therefore, is not drawing up ever longer inventories of different instances of human nature but rather exploring what the diverse instances on that list all have in common. We need to identify the crucial single factor that is so basic that it is independent of different environments yet so sophisticated that it is exclusively human, indeed, according to Wilson, present exclusively in human genes – presumably the 1 per cent of DNA that differentiates us from chimps.

The psychologist Steven Pinker has taken the baton from Wilson and has more recently argued eloquently against the idea that the human mind is a tabula rasa – a blank slate. Despite the brain's plasticity, its large-scale organization is undoubtedly 'genetic', and the small number of genes – still 30,000 at least – are not really as nugatory as all that. This number has turned out to be substantially less than that predicted in Wilson's era, prior to the mapping of the genome; but Pinker points out, pondering how different mutations in genes could lead to large numbers of different permutations and indeed how combinations of genes could yield a very large number of different possibilities, that a 'combinatorial' approach to gene expression paints a far richer landscape. The relatively paltry ratio of genes to brain connections poses no problem if we remember how many different combinations can be derived from a limited number: for example, even a group of just 6 components can generate 720 different combinations!

Indeed, Pinker says, this combinatorial way of looking at the brain might explain how innate behaviours come about. Genes can work together so that different combinations exceed the number of component parts – macro brain-structures could do likewise. In normal circumstances the innate tendency to violence would be offset by another 'natural' brain system that suppressed such behaviour. So the idea that we have natural tendencies is not as scary as all that, since human nature is usually well equipped to prompt a balanced repertoire of action.

But we cannot, as Wilson did, assume that anything as universal as human nature would simply be wired into our genes. Let's briefly revisit the old nature-nurture issue discussed in Chapter 5. Genes are operative not just in the womb but are switched on and off continuously throughout life; moreover they are necessary but far from sufficient factors in determining brain function. So the issue of whether a behaviour is attributable to nature or nurture is quite meaningless.

To recap: a gene triggers the manufacture of a protein, which in turn has a vast range of actions at the molecular level of brain operations, be it the synthesis or removal of a transmitter, or some form of cellular housekeeping or indeed triggering neuronal suicide (apoptosis) – to cite just a few examples. These myriad, basic biochemical mechanisms are exploited in different ways and at different times in different brain regions, which impacts indirectly, and in ways we still do not really understand, on overall brain function and hence finally on outward behaviours. The gene is activated to start off this process not by a wayward agenda of its own but by local chemicals in the nucleus of the cell.

These chemicals can be released locally within the cell at certain stages of development, but in all cases they are not unleashed in a vacuum of isolation. Rather, throughout life different agents activate different genes and switch them off again, to manufacture different proteins to meet different requirements. Sure, sometimes these may be the natural innate requirements of development, but as likely as not they will be switched on by a cycle of events that can be traced back to something in the environment outside of the brain itself – within the mother's womb, or internal systems of hormones washing through the body. But it is the endless interaction between the individual and the external world 'out there' that drives a ceaseless configuring and reconfiguring of brain connections, through the switching on and off of genes. So a gene is simply a tool, one cog in the sophisticated biochemical machinery that translates each influence from an external environment into a physical shift in the pattern of brain connections.

It is therefore impossible to label an emergent behaviour with a nature or nurture provenance. True, there are the twin studies we spoke of earlier, which have found similar dispositions in identical twins reared apart. Some point to an eerie similarity in such cases. However, where this is the case the similarity cannot necessarily be specifically located in precise genes – for example, if both twins had a similar predilection for choosing polka-dot ties, it would not *prove* that there is a 'gene for' love of polka-dot ties. It would be more accurate to say instead that given identical genomes and exposure to similar, usually middle-class environments, similar quirks and habits may emerge in a set of identical twins. We would need to factor out the environment altogether – so far there have been no cases, say, of one twin brought up in New York and one in the Amazonian rain-forest – or we would need to know the exact extent to which any happenstances in their domestic lives or cultural environments really contributed to the entire global mindset of each twin.

A far more interesting question, in any case, is whether or not that quirky common trait, say affection for polka-dot ties, would qualify as 'human nature' simply because it has a strong 'genetic' component. After all, a shared liking for wearing polka-dot ties is remarkable, surely, precisely because not all of us have that trait. So what is human nature, in biological terms? Obviously it is an umbrella term for certain types of behaviour, but one that needs a tighter definition than merely that of having a strong genetic element. Genes, because they are part of the building blocks of the brain, are clearly involved in any behaviours that are attributable to the generic aspects of the human brain – but they are also, by definition, part of what constitutes the non-generic aspects of an individual. Genes, then, will not help us understand the quintessential quality of human nature. Let's look to an older approach.

Way before the famous double helix of DNA was discovered, and the subsequent revolution in gene understanding and technology transformed biomedical science, Sigmund Freud worked out a framework for making sense of behaviours that were all too human. Unable to refer to maps of the human genome and appropriately discouraged from incursions into the physical brain by the primitive neurology and essentially non-existent neurosurgery of his day, Freud had to make do with abstract concepts. Nonetheless, as we all know, his theoretical structure of the generic aspects of the human mind persists to this day: a primitive and universal tendency, common to all animals, towards destruction and creation – the 'id' – the expression of which is channelled by an 'ego' and kept in check by a moralistic 'superego'. It is the ego and superego that distinguishes humans from other animals, but all our behaviours could be stripped down and interpreted as driven by the primitive urges emanating from the id.

This notion was progressed in the early decades of the 20th century by the zoologist Konrad Lorenz, who saw everything we do in terms of the discharge of basic drives. To a certain extent this view of behaviour as 'drive-discharge' makes sense: you are hungry and initiate behaviour to reduce that drive, similarly for sexual desire or sleepiness. But these behaviours are common to all animals, and the fact that we suppress them or devise indirect means of attaining them does not in itself explain the essence of human nature. After all, dogs, cats and certainly primates engage in complex and indirect strategies to achieve a desired endpoint. Again, what is so special about us humans?

If it is not our emotions, or our genes or our basic drives, then where to turn next? Rather than extrapolate from biology perhaps it would be better to work in the opposite direction: let's try to find a good social encapsulation of the non-animalistic yet pan-cultural essence of human nature, and then see how it might be accommodated 'scientifically' within the physical brain.

The seven deadly sins are all exclusively human, and yet the whole point of their biblical catalogue is that they are universal to all times and cultures. But there is something about the list that distinguishes the pursuits of avarice, lust, gluttony, sloth, anger, envy and pride from the universal animalistic drive-reduction behaviours of sex, eating and sleeping. Perhaps the essence of the sins, why they are just that, is the element of excess – too much sex, too much food, too much sleep, too much aggression, beyond what is physiologically needed to sate the natural drives for survival. As for avarice, envy and pride, they are even further removed from crude biological drive reduction but they also evoke excess, within the individual and, more particularly, how that individual fits within the context of their society – their status.

We might overeat because we seek solace for not being loved or fully appreciated; excessive numbers of sexual partners may bring high status among our peers; hoarding of money may reflect the desire to seem important through wealth; resentment of someone perceived to have higher status than oneself is dependent on certain pan-cultural values; those same values, if promoted excessively in the self, are labelled pride; laziness is judged to be so only when certain expectations of one's behaviour are not met; finally anger, contrasted with more straightforward animal-like aggression, implies inappropriate excess, with a causation that is arbitrary, dependent on more ambiguous cultural or even personal values.

Obviously, this is an extremely limited list of examples of phenomena that could be realized in many different ways, as well as being, in psychiatric terms, probably extremely naive and far from accurate; but nonetheless it shows how the seven deadly sins, capturing as they do the universality yet uniqueness of human nature, can be couched in terms of not just certain cultures and values but cultures and values *in general*. Human nature is not as biologically basic as mere drive reduction, or as highly personal as liking specifically polka-dot ties. Rather, human nature is a generalized umbrella term for behaviours that depend on status and social values. The cultures and values may vary, but the behaviours do not. The really important distinction to grasp is that diverse cultures and values determine when and how the behaviours manifest themselves, but the behaviours themselves are recognizable as uniform, above and beyond the particular context in which they occur. Now let's take this idea back to the physical brain.

First we need to go back in time. The archaeologist Steven Mithen has questioned what might have accounted for the sudden explosion, some 100,000 years ago, of intelligence in *Homo sapiens*, and the subsequent development of cave art, language and a host of other exclusively human-type 'cultural' phenomena. He suggests that language is just an example of a wider ability to 'think metaphorically', to see something in terms of something else. My favourite example from his book *The Prehistory of the Mind* is that of a tooth, discarded by some animal, lying in one's prehistoric path. A chimp is able to recognize the object, in isolation, as a tooth. However, humans can go one better and see it not just literally as the sharp white thing that is normally in the mouth but also metaphorically, as part of a necklace. Moreover, that necklace could then be worn as a symbol of status. Never ever do chimps – dextrous and organized in social hierarchies though they may be – sport necklaces as symbols of their status. We are pretty much unique, suggests Mithen, in seeing the world in this way, symbolically as well as literally. Language would be but one example – a particularly powerful one – of our wider human skill in the use of symbols.

So this 'metaphorical' ability of humans could be general enough to be universal, and indeed to encompass humanoid skills like language and art, yet at the same time be sophisticated and specialized enough to be clearly something beyond our basic animal instincts for drive reduction. Perhaps we could think of this broad-spectrum but very human tendency as an updated version of Freud's original ego.

But is this ego not, after all, genetic? Yes, in the sense that it is a feature of the generic human brain, but no, in that it would not be reducible to any one set of specialized genes that do not at the same time underlie the structure and function of the entire brain. The critical issue, then, is not genes per se, any more than transmitters or receptors or neuronal circuits; instead we should ask how all the biochemical baggage assembles into a cohesive operational system that not only coordinates the human body but also amounts to a human mind capable of living a life based on symbols. If human nature is no more and no less than our individualism, how is it grounded in the brain?

In the middle of the 20th century the biologist Paul MacLean effectively attempted a neuroanatomical description of human nature, although he didn't articulate his goal in quite that way. With shades of the Freudian compartmentalization of id, ego and superego, MacLean pointed to the most primitive part of the brain, the brainstem, as the source of our most basic 'reptilian' behaviours, what Freud would have labelled the province of the id. This reptilian aspect of our make-up is, MacLean suggested, kept in check by an overlying system, which he identified as the anatomically enfolding limbic system; in turn, this 'mammalian' brain is kept in check by the cortex, the outer layer of the brain, which increases in surface area, and hence is more crinkled, as the brain increases in sophistication. There are easy parallels between the 'neomammalian' brain of the cortex and Freud's superego. Although this is now regarded as simplistic and naive in terms of anatomical localization of function, MacLean made a nonetheless admirable attempt to explain the then recent behaviours of the crowds at the Nuremberg rallies in biological terms, in terms of human nature. His idea was this: what went wrong was that 'reptilian' crowd behaviour escaped from the normal constraints of the limbic system and was unleashed inappropriately.

But we have seen that there is more to human nature than a simple reversion to being a reptile. The hallmark of the quintessentially human brain is the ability to see things in terms of other things, metaphorically or symbolically. The seven deadly sins would be of no account on a desert island; they are relevant only when human beings start living in a group, a group with certain cultural values, when status counts. Conspicuous consumption, sexual popularity, having more money than anyone else and so on are all tied in with how you the individual are perceived within the context of your culture, by others as well as by yourself.

Central to human nature, therefore, is the concept of the self, and that self will be defined and evaluated in the context of the society in which it lives. A small child feels pleasure from direct stimulation of the senses by the sun or the paddling pool, or the taste of chocolate, or bring rocked to sleep; but if his or her parents win the lottery, it will 'mean' little until he or she has learnt the value systems of the culture – more generally the highly complex, nested series of connections which will permit her to 'understand' the significance of what has happened. The lines from *Macbeth* spouted by my 3-year-old brother were, literally, meaning-less sounds until he was able to attach associations to words like 'death' and, even more advanced, to grasp the notion of an extinguished candle as a metaphor for the end of life. The more the child sees things in terms of other things, the more deeply they 'understand'.

The formation of associations through individual experience – the personalization of the brain – is, we saw earlier, a talent of human beings par excellence. Because the rapid and highly adaptable formation of connections between neurons encompasses language skills, we are able to project ourselves away from the sensory present of the infant into a past and a future conjured up in our minds, and even describable in words or pictorial symbols to others. Above all, as this ability enables us to acquire a sense of the most abstract notion of all, the self, as we develop, we become self-conscious.

We have seen that connections between neurons reflect individual experiences in a nested assembly; the more significant a person or object is, the more associations it will trigger and vice versa. This personalization of the brain, 'the mind', as defined in Chapter 6, could perhaps be as easily viewed as the 'ego'. Once we are self-conscious we are aware of ourselves as separate entities distinct from everything else in the outside world. And just as we can define a tooth as part of a potential necklace, so we will define ourselves in terms of other things too. This process of forming associations, of generating a significance to ourselves, will extend into our lives in whatever ecological niche we find ourselves; we need to be seen in terms of other things, be it prizes, family or possessions. The more associations we can trigger in this way, the more status we have, the more significant and important we are.

This human tendency to see things in terms of other things, to seek 'meaning' to our existence, is particularly relevant as we extend our unique cognitive powers to contemplate the future: by reflecting on what has happened, and what we have been told has happened in the past, we reach the inevitable conclusion that we are going to die. In previous centuries our reaction was to bequeath ourselves immortality: the Renaissance version of the ego might arguably have been the soul. In these more secular times a greater premium has been placed on immortality through our children, or on being so important that in some way at least one's achievements endure. And it is of course this personalized brain, this mind, this ego, this sense of self, that is now under threat from the kinds of new and pervasive influences that we have seen might be starting to dominate in this century.

The basic thesis of this book is that new technologies will have an all-powerful and unprecedented influence on our highly impressionable neuronal connections since, for the first time ever, they will be the sole source of all experience. The prospect becomes even more serious with the potential advent of invasive technologies, which could eventually drive and contrive the configurations of connections directly. But the driving need have no sinister intent – no need to conjure up Big Brother or Huxley's World Controller. Rather, the control could be homeostatic, balanced to sate all physiological drives as and when they surged, and to maintain a sense of non-self-conscious well-being, as though perpetually lying in the sun half asleep after a glass or two of wine.

Previously I have suggested that the basis of pure pleasure is a configuration of the brain such that the active contribution of the personalized connections – the mind – is temporarily non-operational. Small children, lacking extensive connections anyway, are more easily the passive recipients of their sensory input, having a 'sensational' time. Adults, however, turn to extreme measures: drugs that blunt the functioning of our neuronal connections, or such rapid, successive input through our senses from, say, fast-paced sport that no single configuration has time to grow before it is displaced by the start of another. Be it wine, women or song, or the modern analogue, drugs and sex and rock'n'roll, there is a strong premium on sensory, non-associative input; in all cases our brains revert to a simpler, sparser pattern of connections, we recapitulate the 'booming, buzzing confusion' of the infant brain. This state of sensory oblivion, stripped of all cognitive content and bereft of self-consciousness, is probably more like the type of consciousness that most animals experience most of the time. It is to this hedonistic, passive state that the new technologies could be taking us, a state that we enjoy, but that up until now has been only temporary. By incessantly stimulating neuronal connections into certain highly constrained configurations, the new technologies might jeopardize the very existence of human nature, permanently.

But there is an alternative: it could be called the 'public ego'. Let's go back to the Nuremberg crowds that so concerned Paul MacLean. MacLean thought that their behaviour was reptilian – perhaps analogous to what we now recognize as the blind aggression of road rage or a *crime passionnel* or, on a more domestic level, a temper tantrum, where you 'see red': the atavistic urge to destroy, unleashed. Yet although it is true that the Nuremberg crowds behaved in a collective fashion they were *not* directionless. Like the football hooligans of today, the swaggering and shouting crowd is *not* mindless and reptilian: it is all too horribly human. So what if there is, after all, a difference between road rage and the Nuremberg rallies – what if MacLean was wrong?

In *Lord of the Flies*, William Golding captures the collective madness of a group of schoolboys marooned on a desert island; they start as miniature upper-middle-class Englishmen, and end up painting their bodies and killing the weak outsider. Golding's masterpiece was published in 1954, in the post-war climate, and was intended as an illustration of the fragility of civilized society. As he observed in his essay 'On the Nature of Man' (1965): 'My book was to say: you think that now the war is over and an evil thing destroyed, you are safe because you are naturally kind and decent. But I know why the thing rose in Germany. I know it could happen in any country. It could happen here.'

Golding chillingly captured the behaviour of a collective mass of civilized individuals who have in some way abrogated their own private identity. But what they have not done is followed the MacLean version, and ended up in a collective group temper tantrum. No, their aggression is focused and value-related; for the Nazis, for football hooligans and for current fundamentalist fanatics worldwide, the violence is not that of a psychopath, of someone who does not know what they are doing because they are 'out of their mind'.

Instead, there is a very strong sense of group identity and a very clear set of values. There are slogans, an unseen enemy and an abstract cause – all the trappings of adult humans. We saw previously that an element sometimes overlooked in analysis of the Al Qaeda movement is the idea of a great unavenged wrong dating back to the Crusades, when Europeans oppressed the followers of Islam. This highly cognitive narrative of an oppressed minority heroically struggling for its identity against the backdrop of one's own values, which are clearly 'right' whilst those of the prevailing powers are clearly 'wrong', has parallels with certain elements of the Nazi mentality – their romance of Nordic gods and idea of themselves as descendants of the inhabitants of Atlantis, overturning the scheming and powerful Semites. Along more innocent yet comparable lines, fanatical football support is surely all about the triumph of good (your team) over bad (the opposition).

So we are seeing not a loss of identity but quite the opposite: an overemphasis of identity, but one that is collective, a kind of ego that is not private but public. The individual is no longer discernible, sure, but not because he or she has been saturated by an overexuberant id; rather, the ego has become collective. This public ego has all the trappings of its private counterpart – the kinds of features that Wilson listed, and that are exemplified in the deadly sins.

During the hey-day of the Nazis a Berliner, pen-named Sebastian Haffner, documented what was happening to his nation, in an attempt to account for why, against all the odds and rational thought, Hitler was gaining in popularity. Haffner's conclusion was that during the First World War Germans had become used to living as a collective persona – with a public ego. Every day casualty lists were published and battles discussed, with war understandably uppermost in the public consciousness. When it was all over, so Haffner suggests, the Germans found it hard to return to living privately again. In his view the English had pets or gardens whilst their French counterparts had cooking; each individual had enough in their daily lives to give them a sense of identity, a private ego. Germany at that time, he argued, could offer no such attractions. Hence Hitler gained power because he gave the nation back a strong collective sense of identity – he fuelled a public ego.

Haffner's writing is particularly alarming not only because it was written, and never subsequently revised, before the outcome of the Second World War was known – indeed before hostilities even properly started – but also because he describes what happened in Germany between the end of the First World War and the rise of Hitler: the nation became obsessed with sport. Haffner's thesis is that this was yet another way of establishing a public ego. Once again there were battles and victories and a clear set of values and, above all, a strong sense of identity, albeit collective.

Bertrand Russell also alluded to the 'collective passions' of humans as destructive. I think there are today, increasingly, different groups in which individuals are sublimating their private sense of identity to a public one – from the temporary collective public identity that arose

in Britain at the funeral of the Princess of Wales through football
fanaticism to cults and the extreme single-mindedness, literally, of
Al Qaeda.

But if a tendency to embrace a public ego is increasing, how do we
explain the demise of Communism? After all, the very essence of
the original Marxist-Leninist doctrine is the subordination of the
individual for the greater good. The big difference is that, after the
early days of romantic struggle against an oppressor, the Communist
public ego lost its identity as part of a narrative, its role as the underdog.
Where was the glamour, the heroism of the Crusades or of the Nordic
gods? How was the red public ego now significant and meaningful?
Once the element of fighting oppression against all odds was removed,
and the public ego became dominant, what could be said about it other
than that it was in the right? It lacked all character or story, and thus
significance. What was the collective identity and goal? After all, how
could you identify with something that was merely the sum of all of
you . . . Once the memory of 'heroic' leaders, such as Lenin or Stalin,
faded, and as materialism beckoned from capitalist countries, a sense
of identity could be found elsewhere; the appeal of a private ego
became stronger.

We have seen that in the first half of the 20th-century Freud's nephew
Edward Bernays nurtured the development of the private ego by adver-
tising material goods with the claim that they 'stood for' something
special about each individual. Within the culture of the time possessions became symbolic of status and of a particular lifestyle, rather
than being attractive because of their intrinsic qualities alone. In this
cynical way, as Adam Curtis, the producer of *The Century of the Self*,
notes, manufacturing industries, via advertisers, were able to persuade
people to buy products that they did not actually need. But on a more
sinister level, he asserts, Bernays had alighted on a means to prevent
an unleashing of unbridled, basic drives.

My own view is that Bernays was right to fear an alternative to the
private ego, but this alternative was not an unleashed id, or a reptilian
force, or a collective road rage or a perpetual rave – rather, the rise of
a collective identity, a public ego. The ego, be it private or public, is a
bulwark against death, but when it is collective there need no longer
be significance for the life of any one individual – witness the trend of

suicide bombings we discussed in the previous chapter. Moreover, this public ego thrives on some kind of 'meaning' or storyline, as does its private counterpart. The most obvious storyline, to most readily establish significance for a public ego, is struggle – simply establishing one identity as more important than any other. Orwell realized the importance of a struggle to maintain the glamour of a public ego when he wrote in *Nineteen Eighty-Four* of an institutionalized 'Enemy of the People'.

Perhaps the great social upheavals of the 20th century could be seen not as a conflict between Communism and capitalism but more basically *the private versus the public ego*. So far it would seem that the private ego has triumphed, though increasingly threatened by the public ego of extreme terrorist movements. But now we can see that, for the first time ever, the new technologies have the potential to erode the wherewithal of the private ego. This century, then, will witness a competition between movements promoting a public ego and the current trend towards turning the clock back for humanity: perhaps the new technologies could revive a prehistoric stage – before we needed culturally related value systems, a means to be significant as individual people, and where instead we simply lived for the moment, as animals, without recourse to the abstract values and standards of human nature. Participation in the noosphere will not be the same as participation in a cult because, like Communism only very much more so, it will lack a sense of identity. We have seen that it will be a world with no values, no prizes, no goals – no identity. The technological noosphere by which we watch each other, survey each other, will contain no selfishness: it will be the 21st-century version of Communism.

Human nature, then, is in one case obliterated and in the other transformed: the single determining factor as to whether it prevails into this century or succumbs to subversion or extinction is its inherent robustness, or otherwise. Perhaps the behaviours human nature encompasses have lived out their usefulness. This idea is not as strange as it might seem, if indeed the predominant talent of humans is adaptability and if we are about to enter a context-free, value-free, status-free lifestyle favouring passive responses rather than individual action. The fact that we currently work hard to put ourselves into these 'sensational' situations where we feel 'truly alive' may indicate that, with the need for savannah-survival-strategies truly removed, we will slip into a more atavistic state of consciousness where we permanently 'blow' our minds, let ourselves – our private egos – go.

But if our Cro-Magnon brains, with limbic system and crinkly cortex, do mandate a sense of identity, a self-consciousness, then the appeal of football fanaticism and fundamentalist movements will grow as we take an increasingly passive role in a life that is less and less private. Since the corollary of this trend will be, as seems imminent already, large-scale death and suffering if not global nuclear war, then the materially comfortable, anodyne existence in which we lose the essence of our humanity, our human nature, might, amazingly enough, seem more desirable. But are there really no other options?

10

The Future: What are the options?

The technologies that are currently coming of age are not going to go away. Yet the options they appear to offer are stark: they fuel the worst fears of The Technophobe. Both Cynic and Technophile alike scoff that such predictions are sensationalist scaremongering, with no bearing on reality. But there is a fourth group from whom we have not really heard very much as yet, and for whom the options offered by present-day progress are running out. Meet The Vast Majority.

The Vast Majority do not live in places where they can easily appreciate our burgeoning technologies. Since 1980 there has been an increase of more than 50 per cent in younger people in sub-Saharan Africa. Of the swelling world population, half are under twenty-five years old, and by the middle of this century, 90 per cent of all babies will be born in the developing world. Quite soon less than 1 billion people will originate from the industrialized countries: by 2050, for example, there will be three times as many Africans as Europeans.

But this tale of two worlds isn't simply about numbers. Currently, 1.3 billion people are living on less than a dollar a day. In developing countries 4.8 billion lack basic sanitation, whilst one-third of the population has no access to clean water. One billion people in the world are currently illiterate, two-thirds of them women. In many developing countries there is a falling not rising GDP, and in many cases ecological damage is already irreversible: in Brazil, for example, a fifth of the Amazon rainforest has already been destroyed in an attempt to meet the current pressures of having a population of 166 million. What will happen in 2050 when the inhabitants of Brazil will number some 240 million? Such an explosion of humanity deprives developing countries of prosperity, and population growth accelerates global warming, deforestation and loss of groundwater.

As a consequence of this population explosion, more people are living in large urban communities: the number of cities with more than a million inhabitants was 173 in 1990, but that will climb to 368 by 2010. In 1960 only Tokyo and New York could boast more than 10 million inhabitants; by 2015 there will be 26 such 'mega-cities', 22 of which will be in less developed regions. It is in these built-up areas that most people will proliferate most quickly in the next few decades. In 2000 one estimate was that some 2.9 billion were already living in urban areas, about 47 per cent of the world population; but by 2030 this figure will have soared to 60 per cent, and as early as 2007 urban dwellers will exceed their rural counterparts for the first time.

Given the consequences of ecological neglect in the past, as well as shifts towards ever more global markets, life in the country for many is deteriorating to the point of brute survival. As in Britain at the dawn of the Industrial Revolution when to migrate to cities was the only way to survive, we are now globally witnessing desperate people chancing a new, city life; this trend is scarcely surprising. After all, cities offer the opportunities for the most immediate social change and improvement in income. In the future people may well have greater access to schooling, but urban growth will have outpaced employment and services. So there will be no jobs and no decent housing.

Bob Carr, the Premier of New South Wales in Australia, is gloomy as he surveys a planet destined to become more crowded and degraded, hotter, with more desert and less arable land, and with people concentrated into cities: 'On a visit to North-East Asia, I saw this future. The landscape was simple. There were clusters of shoebox-style tower blocks. They were linked by clogged expressways in a flattened, cleared landscape. It was so bleak, so denatured, it could have been a place rebuilt after a nuclear blast. The air was heavy with smog. Acid rain fell from the nation across the ocean. This will be how more people will live in 100 years.'

Now add another factor, presumably linked to this urban migration: the demise of the extended family. In Egypt, for instance, where traditionally such families were the norm, an estimated 84 per cent of all households have now shrunk to the nuclear family. Imagine life, then, as a young citizen of a developing country not too far on into this century. In moving to your current home in the city, your parents cut their close ties with your original cultural roots and clan-like loyalties. Although you may have education, there will be no job. You live in a slum with poor sanitation, but have heard of the amazing technical achievements in industrialized countries. What do you do?

You look towards the industrialized world. By 2050 in developed countries such as Japan, Germany and Italy 40 per cent of the population will be over sixty-five years of age, whilst in the USA even twenty years earlier (2030) 70 per cent will be senior citizens. By 2050 22 per cent of the global population will be over sixty, as opposed to only 10 per cent at the moment, and by the end of the century this figure will have risen to 34 per cent as the industrialized world ages.

But you, the hypothetical young slum-dweller in the developing world, have energy, perhaps some training, no extended family ties, and no work. An obvious, and already much discussed, single solution to the problems of both worlds would be migration to the industrialized countries. This consequence is neither original nor surprising: indeed, it is clearly upon us already. But the ecological, political, financial and sociological implications of a multicultural, enriched world on the one hand and an ever poorer world that has been deserted by the young and able on the other are outside the scope of our discussions here.

Another possibility – albeit remote – is that Western governments succeed in their struggle to establish a streamlined and workable bureaucracy that can differentiate economic migration from asylum-seeking. Some have even suggested gunboats in the Mediterranean, the establishment of a Fortress Europe to keep out The Vast Majority – though no one could seriously buy into this as a viable or acceptable solution. But just imagine such a scenario: behind the bastion of Fortress Europe it is business as usual – and The Vast Majority stay put as the poor relatives of a dwindling and ageing techno-elite far away.

But whether or not large-scale economic migration continues, the problem is the same. If we in the first-world countries continue to develop as we are, with our emails and instant knowledge-base, our Prozac and decriminalized cannabis, our postmenopausal parturition and longer lives, our genetic screening and transplant surgery, our

GameBoy thumbs, Botox injections and *Sex and the City* bachelor lifestyles, then we will have progressively less and less in common with those who have access to none of these things – whoever has been left behind in, or sent back to, the euphemistically labelled developing world. One dire prospect is that we might so diverge as a society that exploitation becomes even more the norm than it is now – even before we factor in the impact of gene therapy, artificial wombs, brain implants, virtual universities and virtual friends, and longer lives crammed with much more leisure yet emptied of any obvious purpose.

The Vast Majority will be increasingly alienated not just socio-economically and culturally but in ways, due to the pace of scientific change, that are so pervasive that we eventually could diverge into 'naturals' versus a gene- and silicon-enriched species. This process of speciation has been a basic feature of the development of lifeforms on Earth throughout evolution. However, as Freeman Dyson writes: 'Speciation in nature occurs with a timescale of the order of a million years. Human speciation pushed by genetic engineering may have a timescale of a thousand years or less. Compared with the slow pace of natural evolution, our technological evolution is like an explosion. We are tearing apart the static world of our ancestors and replacing it with a new world that spins a thousand times faster.'

The interactive, personalized technology that is emerging could, as we have seen, transform humanity on a scale and at a pace far greater than it has ever experienced heretofore; not even the fall of the Roman Empire, the invention of the printing press, the Industrial Revolution, or the large-scale slaughter of the 20th-century world wars has had such an impact. Moreover, in the face of the lifestyle revolutions wrought by bio-, info- and nanotechnologies, The Vast Majority could suffer far more uncompromisingly and completely than the first-world Technophiles, Technophobes and Cynics; they are in danger not only of being disenfranchised from a vastly more comfortable way of life but also of being exploited and abused in ways more sinister, pervasive and cruel than even that witnessed by the worst excesses of the colonialist past.

But there is a third possibility: to harness the ubiquity and accessibility that are the hallmark of the new technologies, to bring the material quality of life in the developing world to a level commensurate with that in first-world countries. Freeman Dyson has suggested that genetically modified trees could generate fuel. This fuel, in combination with solar energy, could be used to power high-tech cottage industries that enable people to remain in the countryside whilst contributing fully to the world economy. Everyone could access all information immediately, independent of location, via the internet; no one need exist in isolation.

Another example of the globalization of technology could be with us as early as 2006, predicts Dick Brass, vice president of technology development at Microsoft: electronic news kiosks that allow people to download newspapers and magazines onto electronic reading devices. By 2010 even this facility will be superseded as lightweight devices with flexible screens and 24-hour batteries render news-on-paper extinct. With increasingly sophisticated hand-held computers, books will be available on screen, and priced at a mere $5, say, compared with $30. One immediate advantage is that the cost of literacy would drop dramatically – every village could have its own electronic library.

And there must be many, many more ways of applying our powerful new technologies, and our still imaginative and proactive minds, to this end. When I served as a judge on the Rolex Awards for Enterprise I was stunned by the ingenuity with which scientists and non-scientists alike have come up with schemes for low-cost electric light, prevention of soil erosion and preservation of vegetables without refrigeration, to cite just a few examples. The increasing access to the new technologies as part of our everyday lives will surely enhance the potential for enterprise and ingenuity enormously – so long as we keep the motivation to let it. Scientists from the developed countries could form a kind of 'Science Peace Corps', and work to harness all the advances we have been surveying to enable the swelling numbers still living a 19th-century, or even medieval, lifestyle, to leapfrog the last hundred or so years and join us in the 21st century.

And there is a less obvious but still more fundamental issue. If we direct technology to these goals, we in the first world might benefit every bit as much as the less technically advanced recipients – perhaps even more so. By taking on stark reality, by tempering the facilities we have in the developed world with the age-old issues of drought, flood, famine, infection, corruption and contraception, we might be able to stave off the more alarming excesses of the new technologies that could corrode our human individuality. In so doing we might improve the lot, the private lives, of those who might otherwise succumb to the glamour of the more public mindset of fundamentalism. The real-life, practical application of new technologies might offer a lifeline out of the dilemma of choosing between subsuming one's identity to a harsh public ego or losing it altogether to the passive sensuality that could rob us of our humanity.

But not everyone will be able or willing to join a Science Peace Corps – indeed it could only ever be formed by a minority of exceptional individuals with innovative minds, huge resources of energy and a deep sense of commitment. For the rest of us, wallowing in techno-luxury, there will be the huge challenge of facing up to the big question of what we want out of life, and ultimately who we are. Initially, each of us will strive, perhaps, to cling to our own personal timetable, our own schedule of goals and achievements, our own life narrative – our own private ego. But . . . 'In the long run, the central problem of any intelligent species is the problem of sanity. We shall be free to choose our values and our purposes. There will be no absolute standards by which to judge one set of values right and another wrong.' Freeman Dyson continues, 'For a society with a technological control of human emotions, addictions to artificial emotional experiences may be fatally easy to induce. A society addicted is this way to dreams and shadows has lost its sanity.'

We shall have to work out how best to set a course between the technophilic Scylla and the technophobic Charybdis. The Techno-phobes, afraid more of what we might do with technology than of technology itself, see salvation in talking the issue through first. Bill Joy suggests that 'If we could agree, as a species, what we wanted, where we were headed, and why, then we would make our future much less dangerous – then we might understand what we can and should relinquish.' He argues that we must find alternative outlets for our creative forces, beyond the culture of perpetual economic growth.

Meanwhile a technophobe website urges, 'It would be good if poli-ticians study physics . . . Often it is only specialists in a narrow field that are usually in the know, and even in this case, there are always discrepancies. In the situation at the turn of the millennium, society and politicians will have to make much more effort to comprehend what is happening, possible perspectives and take steps to prevent potential hazards.'

The idea of everyone simply sitting down and talking is admirable, as is the notion of politicians understanding more about science. But it is not a plausible solution. If we cannot agree, even now, on the pros and cons of some of the nascent technology, say, stem-cell research or GM foods, we would be deluding ourselves to think that we could universally agree on the complex issues that are about to beset society. But rather than picking off, one by one, specific and multifaceted scenarios 'top down', we should work from the bottom up. What single common factor might there be, what single issue is most important in our lives, which we wish to preserve at all costs?

The bottom line of this book is that the private ego is the most precious thing we each have, and it is far more vulnerable now than ever before. It is not an automatic, robust corollary of being born human, but rather depends on the availability of an appropriate environment. We can no longer take such an environment for granted; instead, we need to design and plan for it. Because, at the moment, we each have our own unique manifestos and agendas we shall never reach consensus on all the diverse policy implementations and strategies relating to science and technology. But perhaps we could agree on the ultimate priority to which those policies should be directed – not just the preservation, but also the *celebration* of individuality.

Naive, of course; how easy and attractive to shrug and join the ranks of The Cynics. But then look back at the ground we have covered, starting with domestic gadgets but ending with terrorism, and consider what life might be like if only a fraction of some of the scenarios we envisaged are realized. Remember we are all looking back on the journey with minds shaped by the previous century, but we may soon lose the luxury of cynicism and complacency that that may have engendered. If you are immobilized in the world of 'dreams and shadows', if you are free of pain yet mentally standardized, if you are living principally in a cyber-world or a chemical oblivion, then it may be only of secondary importance whether the rich countries alone undergo this transformation, or the developing world too. And it would no longer matter whether you, or they, had minds of their own. Time could be running out on the luxury of considering any options at all – who knows, we may be the last generation of individuals able, or willing, to have them.

Further Reading

The following will by no means provide exhaustive background to the material covered in the book. However, each discusses specific subjects mentioned in the text in much more detail.

The book has drawn extensively from websites which, due to their instability, cannot be credited precisely.

Aleksander, Igor, *How to Build a Mind*, London: Weidenfeld & Nicolson, 2000.

Ashton, Heather, *Brain Function and Psychotropic Drugs*, Oxford: Oxford University Press, 1992.

Atkins, Peter, *Galileo's Finger: The Ten Great Ideas of Science*, Oxford: Oxford University Press, 2003.

Baker, Robin, *Sex in the Future: Ancient Urges Meet Future Technology*, London: Macmillan, 1999.

Bloom, Floyd E., Flint Beal and David J. Kupfer (eds.), *The Dana Guide to Brain Health*, New York: The Dana Press, 2002.

The Century of the Self, BBC television series, directed by Adam Curtis, first aired on BBC2 in March 2002.

Claxton, Guy, *Wise-Up: The Challenge of Lifelong Learning*, London: Bloomsbury, 1999.

Crick, Francis, *The Astonishing Hypothesis: The Scientific Search for the Soul*, London: Simon & Schuster, 1994.

Davies, Paul, *How to Build a Time Machine*, London: Penguin/Allen Lane, 2001.

Deacon, Terrence W., *The Symbolic Species: The Co-Evolution of Language and the Brain*, London/New York: W. W. Norton & Co., 1997.

Devlin, B., M. Daniels and K. Roeder, 'The heritability of IQ', *Nature* 388 (July 1997), 468–71.

Dyson, Freeman J., *Imagined Worlds*, Cambridge, Mass./London: Harvard University Press, 1997.

—, *The Sun, the Genome, and the Internet: Tools of Scientific Revolutions*, New York: Oxford University Press, 1999.

Evans, P., F. Hucklebridge and A. Clow, *Mind, Immunity and Health: The Science of Psychoneuroimmunology*, London: Free Association Books, 2000.

Frohlich, H., 'The extraordinary dielectric properties of biological materials and the action of enzymes', *Proceedings of the National Academy of Sciences USA* 72 (1975), 4211–15.

Fukuyama, Francis, *Our Post-human Future: Consequences of the Biotechnology Revolution*, London: Profile Books, 2002.

Gershenfeld, Neil A., *When Things Start to Think*, London: Hodder & Stoughton, 1999.

Goldstein, Avram, *Addiction: From Biology to Drug Policy*, 2nd edn, New York: Oxford University Press, 2001.

Gosden, Roger, *Designer Babies: The Brave New World of Reproductive Technology*, London: Gollancz, 1999.

Grand, Steve, *Creation: Life and How to Make It*, London: Weidenfeld & Nicolson, 2000.

Greenfield, Susan, *Journey to the Centers of the Mind: Toward a Science of Consciousness*, New York: W. H. Freeman, 1995.

—, *The Human Brain: A Guided Tour*, London: Weidenfeld & Nicolson, 1997.

—, 'Brain drugs of the future', *British Medical Journal* 317 (1998), 1698–1701.

—, *The Private Life of the Brain*, London: Penguin/Allen Lane, 2000.

Greenough, W. T., J. E. Black and C. S. Wallace, 'Experience and brain development', *Child Development* 58 (1987), 539–59.

Gribbin, John, *The Birth of Time: How We Measured the Age of the Universe*, London: Weidenfeld & Nicolson, 1999.

—, *Science: A History 1546–2001*, London: Penguin/Allen Lane, 2002.

—, et al., *The Future Now: Predicting the 21st Century*, London: Weidenfeld & Nicolson, 1998.

Haffner, Sebastian, *Defying Hitler*, trans. by Oliver Pretzel, London: Weidenfeld & Nicolson, 2000.

Haldane, J. B. S., *Daedalus, or, Science and the Future*, a paper given to the Heretics Society in Cambridge in 1923. First published in London in 1924 by Kegan Paul, Trench, Trubner and Co.; now out of print. The transcribed text is available, however, on the internet.

Hameroff, Stuart, 'Quantum coherence in microtubules: a neural basis for emergent consciousness?', *Journal of Consciousness Studies* 1 (1994), 91–118.

Horgan, John, *The End of Science: Facing the Limits of Knowledge in the Twilight of the Scientific Age*, London: Little Brown, 1996.

—, *The Undiscovered Mind: How the Brain Defies Explanation*, London: Weidenfeld & Nicolson, 1999.

Howe, Michael, *Sense and Nonsense about Hothouse Children: A Practical Guide for Parents and Teachers*, Leicester: BPS Books, 1990.

Huttenlocher, Peter R., *Neural Plasticity: The Effects of Environment on the Development of the Cerebral Cortex*, Cambridge, Mass.: Harvard University Press, 2002.

Huxley, Aldous, *Brave New World*, London: HarperCollins, 1994 (first published by Penguin in 1932).

James, Oliver, *Britain on the Couch: Treating a Low Serotonin Society*, London: Century, 1997.

Joy, Bill, 'Why the future doesn't need us', *Wired* 8.04 (April 2000).

Kaku, Michio, *Visions: How Science will Revolutionize the Twenty-First Century*, Oxford: Oxford University Press, 1998.

Kirkwood, Tom, *Time of Our Lives: The Science of Human Ageing*, London: Weidenfeld & Nicolson, 1999.

Kurzweil, Ray, *The Age of Spiritual Machines: How We will Live, Work and Think in the New Age of Intelligent Machines*, London: Phoenix, 1999.

MacLean, Paul, 'The triune brain, emotion and scientific bias', in F. O. Schmitt (ed.), *The Neurosciences: Second Study Program*, New York: Rockefeller University Press, 1970, pp. 336–49.

Martin, Paul, *The Sickening Mind: Brain, Behaviour, Immunity and Disease*, London: HarperCollins, 1997.

McGee, Glenn (ed.), *The Human Cloning Debate*, 3rd edn, Berkeley, Calif.: Berkeley Hills Books, 2002.

Minsky, Marvin, 'Will robots inherit the Earth?', *Scientific American* 271 (1994), 86–91.

Mithen, Steven J., *The Prehistory of the Mind: A Search for the Origins of Art, Religion and Science*, London: Thames and Hudson, 1996.

Moravec, Hans, *Robot: Mere Machine to Transcendent Mind*, Oxford: Oxford University Press, 1998.

Orwell, George, *Nineteen Eighty-Four*, London: Penguin, 2000 (first published 1949).

Pearson, Ian, www.btinternet.com/~ian.pearson/.

Penrose, Roger, *Shadows of the Mind: A Search for the Missing Science of Consciousness*, Oxford: Oxford University Press, 1994.

Pert, Candace B. and Deepak Chopra, *Molecules of Emotion: Why You Feel the Way You Feel*, New York: Scribner, 1997.

Pesce, Mark, *The Playful World: How Technology is Transforming Our Imagination*, New York: Ballantine Books, 2000.

Pinker, Steven, *How the Mind Works*, London: Penguin/Allen Lane, 1998.

—, *The Blank Slate: Denying Human Nature in Modern Life*, London: Penguin/Allen Lane, 2002.

Rees, Martin, *Our Final Century: The 50/50 Threat to Humanity's Survival*, London: Heinemann, 2003.

Regis, Ed, *Nano!*, London: Bantam Press, 1995.

Ridley, Matt, *Nature via Nurture: The Origin of the Individual*, London: Fourth Estate, 2003.

Rifkin, Jeremy, *The Biotech Century: Harnessing the Gene and Remaking the World*, London: Gollancz, 1998.

Rose, Steven (ed.), *From Brains to Consciousness? Essays on the New Sciences of the Mind*, London: Penguin/Allen Lane, 1998.

Russell, Bertrand, *Icarus, or, The Future of Science*, a response to Haldane's *Daedalus* lecture, published in London in 1924 by Kegan Paul, Trench, Trubner and Co. The text can be found on the internet.

Scientific American, special issue on nanotechnology, September 2001.

Stock, Gregory, *Redesigning Humans: Choosing Our Children's Genes*, London: Profile Books, 2002.

Swain, Harriet (ed.), *The Big Questions in Science*, London: Jonathan Cape, 2002.

Time magazine's V21 reports (available on the internet at www.time.com/time/reportsv21/home.html).

Tutt, Keith, *The Scientist, the Madman, the Thief and Their Lightbulb: The Search for Free Energy*, London: Pocket Books, 2003.

Warwick, Kevin, *QI: The Quest for Intelligence*, London: Piatkus, 2001.

Whalley, Lawrence, *The Ageing Brain*, London: Weidenfeld & Nicolson, 2001.

Wilson, Edward O., *On Human Nature*, London: Penguin, 1995.

Index